DC & Hybrid Micro-Grids

DC & Hybrid Micro-Grids

Editors

Sergio Saponara
Roberto Saletti
Lucian Mihet-Popa

MDPI • Basel • Beijing • Wuhan • Barcelona • Belgrade • Manchester • Tokyo • Cluj • Tianjin

MDPI

Editors

Sergio Saponara
Electronic Engineering
University of Pisa
Pisa
Italy

Roberto Saletti
Electronic Engineering
University of Pisa
Pisa
Italy

Lucian Mihet-Popa
Engineering
Østfold University College
Halden
Norway

Editorial Office
MDPI
St. Alban-Anlage 66
4052 Basel, Switzerland

This is a reprint of articles from the Special Issue published online in the open access journal *Applied Sciences* (ISSN 2076-3417) (available at: www.mdpi.com/journal/applsci/special_issues/dc_hybrid).

For citation purposes, cite each article independently as indicated on the article page online and as indicated below:

LastName, A.A.; LastName, B.B.; LastName, C.C. Article Title. *Journal Name* **Year**, *Volume Number*, Page Range.

ISBN 978-3-0365-1975-3 (Hbk)
ISBN 978-3-0365-1974-6 (PDF)

Contents

About the Editors

Sergio Saponara

Sergio Saponara is an Italian scientist, engineer and entrepreneur, active in the fields of electronics and automotive engineering. He is Full Professor of Electronics at the University of Pisa, Italy, where he is also President of the B.Sc. and M.Sc. degrees in Electronic Engineering. He got his Master and Ph.D. degrees cum laude in Electronic Engineering in 1999 and 2003, respectively. As an entrepreneur, he co-founded the company IngeniArs in 2014, and is the winner of several innovation prizes, such as the H2020 SME Instrument. He founded the Summer School in Enabling Technologies for Industrial IoT in 2016, and is also its director. The organisation was awarded by the IEEE CAS society in 2017 and 2018. In 2020, he was appointed director of the Automotive Electronics and Powertrain Electrifications specialization course for retraining in the field of vehicle electrification for the Vitesco Technologies (Continental) group.

Roberto Saletti

Prof. Roberto Saletti graduated with honors in Electronics Engineering at the University of Pisa, Italy in 1981. He was research scientist of the Italian National Research Council since 1983 to 1992. In 1987 he was visiting scientist at the Cornell University, Ithaca, New York. In 1992 he was nominated associate professor of Electronics at the Faculty of Engineering of the University of Pisa and full professor of Electronics in September 2001. Since 2016 is elected member of the Senate of the University of Pisa. His main research activities are in the fields of design and test of embedded electronic systems for data acquisition and processing and in-vehicle electrification, particularly in battery control. He is Senior Member of the IEEE and member of the Industrial Electronics Society (IES) and the IEEE Instrumentation and Measurements Society (IM). He coauthored more than 150 scientific publications in peer-reviewed international journals and conferences.

Lucian Mihet-Popa

Lucian Mihet-Popa received his Habilitation (2015) and Ph.D. degree (2002) in Electrical Engineering, a Master's degree in Electric Drives and Power Electronics and a Bachelor's degree in Electrical Engineering, from the Politehnica University of Timisoara-RO. Since 2016, he has worked as Full Professor in Energy Technology at Oestfold University College in Norway. Dr. Lucian Mihet-Popa has published more than 150 papers in national and international journals and conference proceedings, and 12 books. Since 2017, he has been a Guest Editor for five Special Issues for the MDPI *Energies* and *Applied Sciences* Journals. Professor Mihet-Popa has participated in more than 15 international grants or projects, such as FP7, EEA, and Horizon 2020, and has been awarded more than 10 national research grants. He is also the head of the Research Lab "Intelligent Control of Energy Conversion and Storage Systems" and one of the Coordinators of the Master Program in Green Energy Technology.

Preface to "DC & Hybrid Micro-Grids"

This book is a printed version of the papers published in the Special Issue "DC & Hybrid Microgrids" of Applied Sciences. This Special Issue, co-organized by the University of Pisa, Italy, and Østfold University College in Norway, has collected nine papers and the editorial, from 28 submitted, with authors from Asia, North America, and Europe. The published articles provide an overview of the most recent research advances in direct current (DC) and hybrid microgrids, exploiting the opportunities offered by the use of renewable energy sources, battery energy storage systems, power converters, innovative control, and energy management strategies.

Sergio Saponara, Roberto Saletti, Lucian Mihet-Popa

Editors

applied sciences

MDPI

Editorial

Recent Trends in DC and Hybrid Microgrids: Opportunities from Renewables Sources, Battery Energy Storages and Bi-Directional Converters

Sergio Saponara [1,*], Roberto Saletti [1] and Lucian Mihet-Popa [2]

1 Department of Information Engineering, University of Pisa, V. Caruso 16, 56122 Pisa, Italy; roberto.saletti@unipi.it
2 Faculty of Engineering, Østfold University College, Kobberslagerstredet 5, 1671 Fredrikstad, Norway; lucian.mihet@hiof.no
* Correspondence: sergio.saponara@unipi.it

Received: 22 March 2020; Accepted: 4 June 2020; Published: 26 June 2020

check for updates

Featured Application: Hybrid microgrids for residential applications with connections to renewable energy sources, electric vehicle recharging stations and the AC energy grid.

Abstract: This editorial manuscript reviews the papers accepted for publication in the Special Issue "DC & Hybrid Microgrids" of *Applied Sciences*. This Special Issue, co-organized by the University of Pisa, Italy and Østfold University College in Norway, has collected nine papers from 25 submitted, with authors from Asia, North America and Europe. The published articles provide an overview of the most recent research advances in direct current (DC) and hybrid microgrids, exploiting the opportunities offered by the use of renewable energy sources, battery energy storage systems, power converters, innovative control and energy management strategies.

Keywords: microgrid; direct current (DC) grid; hybrid microgrid; smart grid; battery energy storage; renewable energy sources (RES); hybrid vehicles (HEV); full-electric vehicles (EV); bidirectional converters

1. Introduction

This editorial analyzes the manuscripts accepted, after a careful peer-reviewed process, for the Special Issue "DC & Hybrid Microgrids" of *Applied Sciences*. The Special Issue has been co-organized by the University of Pisa (Full Professors Sergio Saponara and Roberto Saletti at the Department of Information Engineering), Italy and Østfold University College (Full Professor Lucina Mihet-Popa at the School of Engineering) in Norway.

The nine accepted manuscripts were co-authored by research groups from three continents, consisting of authors from Norway, Denmark, Italy in Europe; India, China, Japan, Vietnam in Asia; and the USA in North America.

As reported in Section 2 of this editorial, the selected papers give an overview of the trends in research and development activities about direct current (DC) and hybrid microgrids. By exploiting the opportunities offered by the use of renewable energy sources, battery energy storage systems, power converters, and innovative control and energy management strategies, new scenarios are opened with the opportunity for industrial exploitation.

2. Recent Trends in DC & Hybrid Microgrids

The Special Issue is characterized by eight original research papers [1–8] and one review paper [9].

The first paper [1] entitled "Research on Virtual Inductive Control Strategy for Direct Current Microgrid with Constant Power Loads", was written by Zhiping Cheng et al., authors from the School of Electrical Engineering, Zhengzhou University, China. Aiming at improving the stability of DC microgrids with constant power loads, the paper presents a novel virtual inductive approach. Indeed, it is known that the negative impedance characteristic of constant power loads will lead to DC bus voltage fluctuations. A simplified circuit model of the system has been obtained for the analysis by modeling the distributed resources. Unlike the existing control strategies, the proposed controller constructs a negative inductance link which helps to counteract the negative effects of the line inductive between the power source and the transmission line. Detailed performance comparisons of the proposed control and virtual capacitance are carried out by means of MATLAB/Simulink simulations. The improved performance of the proposed control strategy has been further validated with several detailed studies and the achieved results demonstrate the feasibility of the proposed approach.

The second paper [2], entitled "Improvement of an Islanding Detection Method Based on the Perturbation Signal in Case of a Multi-Photovoltaic Operation", is by Thanh Son Tran et al., from the Graduate School of Engineering and Science and the Department of Electrical Engineering, Shibaura Institute of Technology, Tokyo, Japan and the Department of Power System, Hanoi University of Science and Technology, Hanoi, Vietnam. The paper deals with the issues of the islanding phenomenon, which is one of the consequences of the emergence and development of microgrids in the power system. A common problem of a multi-distributed generation (particularly when using multiple renewable energy sources, such as the multi-photovoltaic sources addressed in this paper) is the cancellation of the injected signals, which has a significant influence on active islanding detection methods. This paper analyzes this issue by injecting a perturbation signal in a multi-photovoltaic system. A promising solution for the injected signal cancellation was also proposed in the manuscript and validated through mathematical explanations and simulations.

The third paper [3], entitled "Insulation Monitoring Method for DC Systems with Ground Capacitance in Electric Vehicles", was written by Jifei Du et al., from the School of Electrical Engineering, Beijing Jiaotong University, and the National Active Distribution Network Technology Research Center (NANTEC), Beijing Jiaotong University. The paper deals with the issue of the influence of ground capacitance in electric vehicles, so that the voltage of positive and negative electric bridges changes slowly in the traditional unbalanced electric bridge DC insulation monitoring (DC-IM) method. Sampling should be conducted once the voltage of the bridge becomes stable, to calculate the insulation resistances. This fact will inevitably extend the monitoring cycle. The paper proposes a three-point climbing algorithm to reduce the monitoring cycle. The algorithm consists in a three-bridge voltage sampling with equal sampling intervals, to predict the evolution of the bridge voltage curve. The insulation resistances calculated by sampling values will deviate from the actual values because of the presence of sampling errors. The paper also proposes a filter and a correction method to solve this issue. Experimental data are reported to show the influence of different parameters on the results and to make a comparison with the traditional method. When compared to the state of the art, the proposed technique can monitor insulation resistance more quickly and can ensure fixed monitoring cycles under different ground capacitance values, while keeping a similar monitoring accuracy.

The fourth paper [4], entitled "Improved Control of Forest Microgrids with Hybrid Complementary Energy Storage", by Ming Yu et al. is also from China, particularly from the School of Technology, Beijing Forestry University. The paper proposes a hybrid complementary energy storage control method to improve the power quality and the fault ride-through capability of islanded forest microgrids. In the paper, mode-based sectional coordinated control is adopted as the basic control scheme, whereas control of the hybrid energy storage adopts the improved strategy. The hybrid energy storage system contains a battery and a supercapacitor coupled to a wind turbine subsystem. Given the different characteristics of the energy storage units, the adaptive control of battery and supercapacitor are adopted in the paper to smooth the low-frequency power fluctuation in the long term and to suppress the high-frequency component separately. It is to be noted that a predictive control of the converters is adopted to achieve rapid regulation. The wind power unit is also investigated as third energy storage unit. The idea involves utilizing the large rotating kinetic energy of the wind turbine to temporary suppress huge power disturbances and to avoid load shedding. The paper show simulations of the islanded DC microgrid in forest areas, performed with MATLAB/Simulink, to validate the effectiveness of the proposed coordination control with hybrid complementary energy storage. The achieved results show that the transient operation characteristics of the system were effectively enhanced by utilizing the improved control method.

The fifth paper [5], entitled "An AC/DC Distribution Network DG Planning Problem: A Genetic-Ant Colony Hybrid Algorithm Approach", is by Deyang Yin et al. from the School of Electrical Engineering, Southeast University, Nanjing, China and the College of Energy and Electrical Engineering, Hohai University, Nanjing, China. The paper addresses the topic of the planning problem of distributed generators (DG) accessing the AC/DC distribution network. The paper presents a location and volume model of a DG that considers several DG costs, such as operation and maintenance costs, investment costs, system network loss costs, fuel costs, pollution compensation costs, and environmental protection subsidies. Voltage and power constraints are also considered in the model. A hybrid algorithm that combines the ant colony algorithm (ACO) and the genetic algorithm (GA) is proposed to solve the model presented, resulting in a GA-ACO technique. The IEEE-33 node distribution network is taken as example to verify the rationale of the proposed approach and its effectiveness. The simulation results show that the proposed model is aligned with reality and the hybrid algorithm is effective in solving the model, with advantages in both convergence speed and convergence results, if compared to single ACO and/or GA techniques.

The sixth paper [6], entitled "A Stability Preserving Criterion for the Management of DC Microgrids Supplied by a Floating Bus", by D. Bosich et al. is from Italy, from the Department of Engineering and Architecture, University of Trieste and the Dipartimento di Elettronica, Informazione e Bioingegneria, Politecnico di Milano in Milan. One of the most important enabling technologies for the future development of microgrids is DC distribution. This is due to the easier interfacing with many DC components of a modern smart grid, such as batteries, photovoltaic systems, and native DC loads. The large use of controlled power converters suggests the need for a careful analysis of system stability in these power systems, as the stability can be impaired in particular conditions. For example, a destabilizing effect can arise in DC power systems due to the presence of inductor/capacitor (LC) filtering stages installed for power quality requirements. Other stability problems can arise due to high-bandwidth controlled converters that behave as constant power loads (CPLs). This issue is even more critical when the CPL is potentially fed only by the battery, causing the DC bus to float. A valuable method for studying the system stability of DC microgrids feeding CPLs is the Lyapunov theory, which demonstrates how the region of asymptotic stability (RAS) shrinks, as the state of charge of the battery diminishes and the bus voltage decreases. Once the accuracy of the RAS is validated by comparing it to the real basin of attraction (BA) and numerically derived using continuation methods, a smart power management of the CPL is proposed in the paper to preserve the system stability, even in presence of a low bus voltage. As discussed in the paper, a suitably designed criterion for limiting the load power can guarantee the invariance of RAS and BA for each equilibrium point. An electric vehicle was also used in the paper as particular DC microgrid component for evaluating the performance derating given by the power limitation.

The seventh paper [7], entitled "Multidimensional Optimal Droop Control for DC Microgrids in Military Applications", is by K. J. Bunker at al. from the USA, from the Department of Electrical and Computer Engineering, Michigan Technological University in Houghton; the Rocky Mountain Institute, Boulder; and the Department of Mechanical Engineering and Engineering Mechanics, Michigan Technological University. The paper deals with reliability issues of microgrids when they are used for defense applications. Droop control provides a simple option without requiring communication between microgrid components, increasing the control system reliability. As discussed in the paper, traditional droop control does not allow the microgrid to utilize much of the power available from solar renewable sources, such as a photovoltaic resource. The paper shows the application of an optimal multidimensional droop control strategy for a solar resource connected in a microgrid at a military patrol base to solve this issue. The paper reports simulation and hardware-in-the-loop experiments of a proof-of-concept microgrid to show that much more power from the solar resource can be utilized, while maintaining the system's bus voltage around a nominal value, and still avoiding the need for communication between the various components.

The last original research paper of the special issue [8], entitled "Energy Management Strategy for Rural Communities' DC Micro Grid Power System Structure with Maximum Penetration of Renewable Energy Sources", written by M. Gunasekaran et al., it is a collaborative work by three research groups from the Department of Electrical and Electronics Engineering, SRM University, Chennai, India; the Faculty of Engineering, Østfold University College, Norway; and the Department of Energy Technology, Aalborg University, Denmark. Regarding classic AC grid with power distribution to many DC loads, DC microgrids are suitable to provide high efficiency, consistency, reliability, and load sharing performance, particularly when interconnected to DC renewable and storage sources. The main control objective for any DC microgrid is providing proper load–power balancing based on the distributed generator sources. Due to the intermittent nature of renewable energy sources, batteries play an important role in load–power balancing in a DC microgrid. State-of-the-art energy management strategies may be able to meet the load demand, but often such techniques are not suitable for rural communities. To solve this issue, the paper proposes an energy management strategy (EMS) for a DC microgrid to supply power to rural communities with solar, wind, fuel cell, and batteries as input sources. The proposed EMS performs the load–power balancing between each source (renewable and storage) in a DC microgrid for dynamic load variations. The EMS proposed in the paper, which manages two battery sources, is capable of achieving load–power balancing using renewable energy sources with less consumption than non-conventional energy sources such as a diesel generator. The performance of the system is analyzed based on the different operating conditions of the input sources and verified by means of both simulation models and a laboratory real-time DC microgrid experimental setup.

Finally, the review paper [9], entitled "Hybrid Microgrids Exploiting Renewables Sources, Battery Energy Storages, and Bi-Directional Converters", is co-authored by the Special Issue guest editors from University of Pisa, Italy and Østfold University College, Norway. With reference to a new and improved energy grid, the guest editors' review paper analyzes trends in renewable energy sources, power converters, and control strategies, as well as battery energy storage technologies and the relevant issues in battery charging and monitoring. The paper also proposes an alternative microgrid architecture overcoming the lack of flexibility of the classic energy grid. By mixing DC and AC sources, the proposed hybrid microgrid stands as an alternative solution where the use of bidirectional electric vehicle chargers creates a microgrid that directly interconnects all the partner nodes with bi-directional energy flows. This model is further sustained by the new products emerging in the market, since new solar inverters are appearing in which a local energy storage system for the renewable energy sources is available. Therefore, the power flow from/towards the renewable energy sources becomes bidirectional, with improved flexibility and efficiency.

Author Contributions: Research: S.S., R.S., and L.M.-P. Writing: S.S., R.S., and L.M.-P. All authors have read and agreed to the published version of the manuscript.

Funding: This work was partially supported by Tuscany Region under "Progetto SUMA" and partially supported by CrossLab project, University of Pisa, funded by MIUR "Department of Excellence" program.

Acknowledgments: Discussions with K.J., L.C., W.X. from MIT, US under the MIT-UNIPI seed fund program are gratefully acknowledged.

Conflicts of Interest: The authors declare no conflict of interest.

References

1. Cheng, Z.; Gong, M.; Gao, J.; Li, Z.; Si, J. Research on Virtual Inductive Control Strategy for Direct Current Microgrid with Constant Power Loads. *Appl. Sci.* **2019**, *9*, 4449. [CrossRef]
2. Tran, T.S.; Nguyen-Duc, T.; Fujita, G. The Improvement of an Islanding Detection Method Based on the Perturbation Signal in Case of a Multi-Photovoltaic Operation. *Appl. Sci.* **2019**, *9*, 4054. [CrossRef]
3. Du, D.; Zheng, X.; Yan, Y.; Zhao, H.; Zeng, X.; Li, L. Insulation Monitoring Method for DC Systems with Ground Capacitance in Electric Vehicles. *Appl. Sci.* **2019**, *9*, 2607. [CrossRef]
4. Yu, M.; Zhang, J.; Liu, H. Improved Control of Forest Microgrids with Hybrid Complementary Energy Storage. *Appl. Sci.* **2019**, *9*, 2523. [CrossRef]
5. Yin, D.; Mei, F.; Zheng, J. An AC/DC Distribution Network DG Planning Problem: A Genetic-Ant Colony Hybrid Algorithm Approach. *Appl. Sci.* **2019**, *9*, 1212. [CrossRef]
6. Bosich, D.; Vicenzutti, A.; Grillo, S.; Sulligoi, G. A Stability Preserving Criterion for the Management of DC Microgrids Supplied by a Floating Bus. *Appl. Sci.* **2018**, *8*, 2102. [CrossRef]
7. Bunker, K.J.; Cook, M.D.; Weaver, W.W.; Parker, G. Multidimensional Optimal Droop Control for DC Microgrids in Military Applications. *Appl. Sci.* **2018**, *8*, 1966. [CrossRef]
8. Gunasekaran, M.; Ismail, H.M.; Chokkalingam, B.; Mihet-Popa, L.; Padmanaban, S. Energy Management Strategy for Rural Communities' DC Micro Grid Power System Structure with Maximum Penetration of Renewable Energy Sources. *Appl. Sci.* **2018**, *8*, 585. [CrossRef]
9. Saponara, S.; Saletti, R.; Mihet-Popa, L. Hybrid Micro-Grids Exploiting Renewables Sources, Battery Energy Storages, and Bi-Directional Converters. *Appl. Sci.* **2019**, *9*, 4973. [CrossRef]

applied sciences

MDPI

Review

Hybrid Micro-Grids Exploiting Renewables Sources, Battery Energy Storages, and Bi-Directional Converters

Sergio Saponara [1,*], Roberto Saletti [1] and Lucian Mihet-Popa [2]

[1] Dipartimento di Ingegneria della Informazione, Università di Pisa, via G. Caruso 16, 56122 Pisa, Italy; roberto.saletti@unipi.it
[2] Faculty of Engineering, Østfold University College, Kobberslagerstredet5, 1671 Fredrikstad, Norway; lucian.mihet@hiof.no
* Correspondence: sergio.saponara@unipi.it

Received: 28 September 2019; Accepted: 9 November 2019; Published: 19 November 2019

Featured Application: Hybrid micro-grids for residential applications with connections to renewable energy sources, electric vehicle recharging stations and the AC energy grid.

Abstract: This paper analyzes trends in renewable-energy-sources (RES), power converters, and control strategies, as well as battery energy storage and the relevant issues in battery charging and monitoring, with reference to a new and improved energy grid. An alternative micro-grid architecture that overcomes the lack of flexibility of the classic energy grid is then described. By mixing DC and AC sources, the hybrid micro-grid proposes an alternative architecture where the use of bi-directional electric vehicle chargers creates a micro-grid that directly interconnects all the partner nodes with bi-directional energy flows. The micro-grid nodes are the main grid, the RES and the energy storage systems, both, on-board the vehicle and inside the micro-grid structure. This model is further sustained by the new products emerging in the market, since new solar inverters are appearing, where a local energy storage for the RES is available. Therefore, the power flow from/towards the RES becomes bi-directional with improved flexibility and efficiency.

Keywords: micro-grid; hybrid micro-grid; smart grid; battery energy storage; renewable energy sources (RES); hybrid vehicles (HEV); full-electric vehicles (EV); Bidirectional converters

1. Introduction

The need to reduce CO_2 and greenhouse emissions, and the aim of achieving zero-energy or near-zero-energy buildings, have sustained the research, development, and adoption of Renewable Energy Sources (RES) for buildings and for electrified propulsion of vehicles [1–5].

As far as buildings are concerned, either, public (e.g., offices, schools ...) or private photovoltaic (PV) sources have mainly been used. On the electric vehicle side, plug-in hybrid electric vehicles (PHEV) or full electric vehicles (PEV) are gaining market shares [1], mainly using battery-based energy storage rather than using fuel cells and hydrogen tanks.

Increasing the amount of low-carbon energy production means a larger diffusion of small-scale generation sites are based on RES and located at the distribution grids. In order to take full advantage of available RES, Energy Storage Systems (ESSs) are also necessary. This results in a scenario full of complex systems with numerous active elements that need to properly be managed and controlled. Every level of the above described scenario must be considered, starting from the individual power converters and their control of new business schemes that are necessary to fulfill.

The concept of micro-grids with classification schemes as a solution for reliable integration of Distributed Energy Resources (DERs), including energy storage systems and controllable loads, can be found in various literature, such as in [6–11]. Although the common configuration of micro-grids is AC, a huge interest in DC micro-grids has been shown for some advantages, such as no reactive power or synchronization need, and the increasing number of available smart DC devices and DC loads. However, hybrid micro-grid combines the best aspects of the two configurations.

Hybrid micro-grids combine RESs with diesel gen-sets and energy storage technologies, mainly used as backup systems to deliver clean, cost-effective electricity to remote locations, with limited or no access to reliable utility power systems. These hybrid systems, based on RES, can contain several parallel-DERs, which are able to operate in both islanded and grid-connected modes [12–15]. These configurations have the advantages of improving environmental performances as well as resiliency and reliability, providing uninterruptible power supply to critical loads while meeting customer satisfaction at a reduced cost by providing autonomy, safety and security for all actors involved. Advanced hybrid micro-grids are also able to balance electrical demand of RES, scheduling the dispatch of resources and preserving the grid reliability, managing the integration and interoperability of complex configurations of distributed generation, storage and controllable loads, demand response and EMS (energy management systems) [16–18].

In order to control the load sharing among AC and DC sources, hybrid micro-grids operation requires stable and appropriate power management strategies, which are mainly based on droop control. There are valid conventional solutions for the power management of different generating resources. The emerging number of distributed generation (DG) nodes can be handled by aggregators considering the energy market aspects. Considering that most of the inverter-based resources are located at the distribution network, the capacity of the local grid may be a bottleneck.

It is well known that the power electronic converters play an important role in the hybrid micro-grids in the context of conversion, generation, distribution and power flow control as well as for energy management and energy efficiency issues and integration capabilities [19]. Recently, power switches manufactured with silicon-carbide (SiC) and gallium nitride (GaN) transistors can further increase the efficiency and power control stability, reducing at the same time the size of the power electronic modules and systems [20–23].

Vehicle to Grid (V2G) is considered an innovative and disruptive solution that could strengthen the development and wide spreading of the energy management in micro-grids. The possibility for aggregated car owners to provide balance services to the grid will give a benefit (and revenue stream) that allows the gaps to be filled with conventional transport.

The grid organization typically follows a scheme like in Figure 1, where there is no direct energy flow or communication among the RES, e.g., PV panels with relevant solar inverter mounted on the roof of the building, and the energy storage system installed on-board the vehicle. All communications and energy flows are typically implemented between the RES sub-system and the main grid (managed by a national electricity provider, e.g., ENEL in Italy) and between the main grid and the vehicle to be recharged. The energy flows in Figure 1 are mainly unidirectional: From PV source towards the grid node, due to a solar inverter (DC/AC energy conversion); and from the main grid to the energy storage (battery pack) on-board the vehicle for EV recharging, due to an AC/DC power conversion.

This classic approach has some limits that have not yet been overcome. RESs are intermittent in nature. For example, the produced energy is limited at night-time, in case of cloudy days or during less sunny seasons, when using PV sources. Moreover, when a recharged car is parked (e.g., during office time), the energy stored on-board cannot be used as a source to fulfill the energy requirements of other loads. The lack of flexibility of the classic energy grid in Figure 1 can be overcome by a new energy grid concept, that is described in this survey paper in Section 4. By mixing DC and AC sources, the hybrid micro-grid in Section 4 proposes an alternative architecture where the use of bi-directional EV chargers may allow a micro-grid to be created by directly interconnecting with bi-directional energy flows all the nodes: The main grid node, RES node, energy storage nodes, both on-board the vehicle, and inside the micro-grid structure. This model is further sustained by the new market products, since new solar inverters are appearing [2], e.g., from ABB, where a local energy storage for the RES is foreseen, and hence, the power flow from/towards the RES becomes bi-directional.

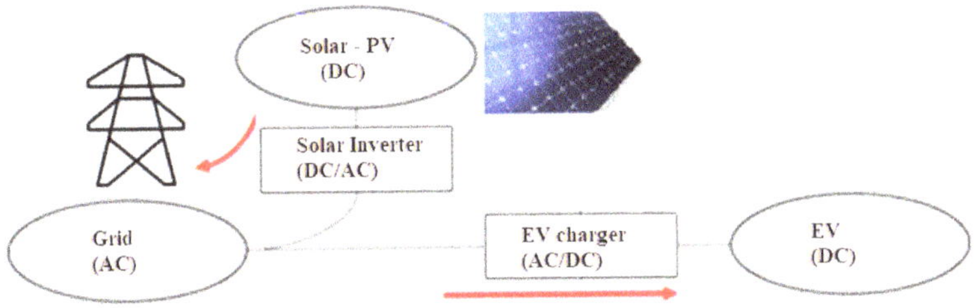

Figure 1. Classic local grid where RES and EV charger nodes are only connected to the main grid and with unidirectional energy flows. RES: renewable-energy-sources; EV: full-electric vehicles; PV: photovoltaic.

Hereafter, Section 2 reviews trends in RES, power converters and controllers. Section 3 analyzes trends in battery energy storage and the relevant issues in battery management and charging. Section 4 presents an alternative micro-grid architecture. Conclusions are finally drawn in Section 5.

2. Renewable Energy Source Trends for Micro-Grids

2.1. Renewable Energy and Micro-Grids

Renewable energy will have the fastest growth in the electricity sector for the next five to six years, and is the central stage of the transition to less CO_2 emissions and more sustainable energy. Renewables like wind power and solar power have grown very fast in the past 10 years, particularly because of their cost reduction, which has a 50% reduction target of 2030. Even if the electricity generated by Renewable Energy Sources-RES represents is only one fifth of the global energy consumption, the roles of renewables in the transport and heating sectors still remain a huge challenge in sustaining the energy transition. International Energy Agency-IEA [24] estimated in 2018 that the variable RES, such as wind and PV, but also hydropower and bioenergy, will have a significant power generation over the next five years, providing 30% of power demand in 2023 from 24% which was in 2017, as shown in Figure 2. This means that the power capacity expansion reaching at around 70% of global energy generation will grow in the next five years. Distributed solar PV is counting for almost half of total solar PV grows over 2019–2024 [25]. The shares of RES in electricity, heat and transport for 2017, with an estimation for 2023 can also be seen in Figure 2.

The modern micro-grid concepts incorporate multiple DERs, power electronic converters, and different control strategies, such as active power versus frequency, or reactive power versus voltage. The final objective is to remove challenges to smart grids integration, remove reliance on high-speed communication and peer-to-peer architectures, as well as create a plug-and-play reliable system.

2.2. Power Electronic Converters Used in Micro-Grids

Micro-grids deal with a wide range of topics, such as power electronics, power systems, RES generation and storage, and ICT (Information and Communication Technology). The conventional utility grid works in a passive mode absorbing energy from the network and delivers it to customers. This approach is well-known, well-developed, and presented in many research papers, but the modern micro-grids/smart grids need state-of-the-art technology, including a bi-directional power flow and big-data processing/cloud computing. Modern micro-grid systems should provide more flexibility, reliability, sustainability, security, and two-way communication services, as shown in Figure 3. In particular, the integration of RES, EVs and DGs into power systems is achieved in an efficient way in micro-grid systems, based on power electronics with wide band-gap devices (IGBT (Insulated Gate Bipolar Transistor)s and MOSFET (Metal-Oxide-Semiconductor Field-Effect Transistor)s), such as gallium nitride on silicon (GaN-on-Si) technology.

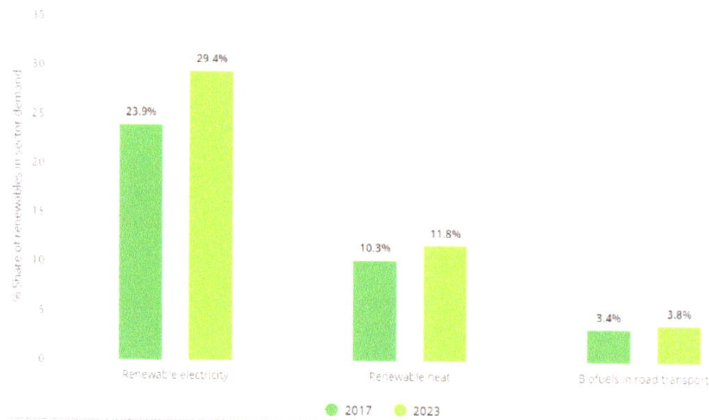

Figure 2. A comparison between the shares of RES in electricity, heat and transport for 2017 and 2023. RES: Renewable Energy Sources.

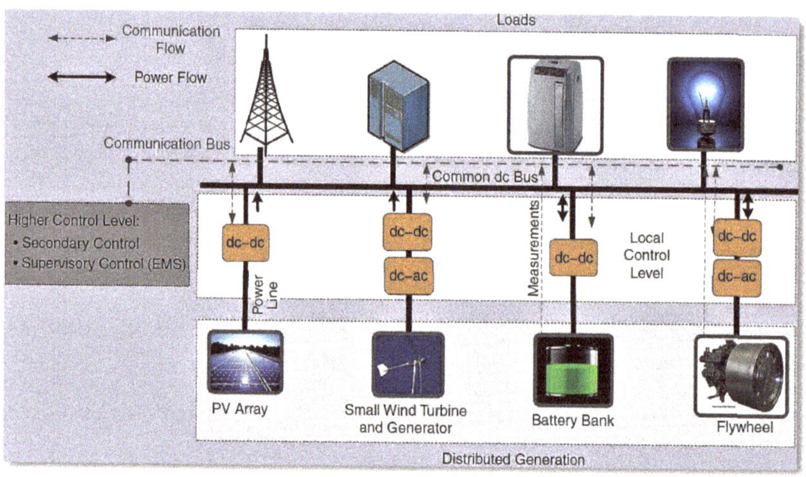

Figure 3. Power electronic converters used in micro-grids.

Distributed Energy Resources (DER) components used in micro-grids, including PV systems, Wind Turbines (WTs), ESSs, EVs, require power electronics interfaces like DC/AC inverters, bi-directional DC/DC converters or AC/DC/AC converters, as can also be seen in Figure 3. A micro-grid can include many converter topologies to interface DG units and loads. For instance, the PV array in Figure 3 connects a number of PV modules to a one quadrant DC-DC converter interfacing the PV-generated power to the load via the common DC bus, when operates in stand-alone hybrid mode, together with a battery bank that is connected to the common DC bus via a bidirectional DC-DC converter. The wind turbine generator and the flywheel system are connected to the common DC bus through a DC-AC inverter and a DC-DC converter. In this case, the DC-DC converter connected to the DC-link/DC-bus can serve as a storage or as an electrical brake, for wind turbine during wind gusts.

DC/AC inverters play an important role in frequency and voltage control, while AC/DC converters can isolate the micro-grid from the utility and properly match it with the DC loads in residential applications [9,26,27]. Inverters for utility connection can be broadly classified into two types: Switch-mode single-phase inverters and switch-mode three-phase inverters. The detection of islanding is much easier in a three-phase than a single-phase inverter. Although, inverters that are rated at a power below 5 kW are mostly connected to single-phase networks.

The most important power interfaces between DG sources and loads (DG inverters) employed voltage source inverters (VSI), which can be classified in two categories [28–30]: current-controlled inverter (CCI) and voltage-controlled inverter (VCI). CCI could also be used as an active power filter to reduce/mitigate the THD (Total Harmonic Distortion) factor of the converter, improving in the same time the power quality of the hybrid micro-grid. Instead, VCIs serve for supplying the active and reactive powers as well as to provide voltage and frequency regulation for islanding operation of the micro-grid.

Hybrid micro-grids have the benefits of both AC and DC micro-grids in supplying conventional AC loads and AC sources on AC side (AC bus) and DC loads and DC sources on DC side (DC bus), in controlling the corresponding individual micro-grid (sub-grid) and then in coordinating and interconnecting the AC with DC micro-grids through an interlinking converter [31–34], in which the DC-side is dynamically decoupled from the AC-side. Interlinking converters play a key role in hybrid micro-grids regarding the grid-tied and islanding operations and the seamless transition and resilience to main grid perturbation (LVRT-Low Voltage Ride Through capability), active harmonic filtering capability and power-flow control.

Each converter type requires a proper control strategy/algorithm that is selected among numerous methods, based on specific requirements and applications. The control algorithms could be classified according to their loop either, in feedback or in feed-forward structure. For instance, the controller of DC/AC switch-mode inverters can contain six parts: Maximum power controller, ESS charge controller, voltage controller, harmonic controller, RMS voltage controller, and current waveform generator.

Many DER components provide or generate a DC current, which is part of an internal DC micro-grid to avoid losses and permits plug-and-play capabilities. The hybrid energy systems involve a well-optimized energy control and management infrastructure to achieve the highest possible energy level.

Moreover, control strategies and monitoring capabilities, automatic control, and configuration of the grid, and the active involvement of the consumers in energy production extend the importance of the micro/smart grids. All the advantages of smart grids can be achieved by power electronic converter integration and ICT technologies with the grid [10,26,27].

2.3. Control Strategies Used in Hybrid Micro-Grids

Two particular and opposite approaches are used with respect to the power system's control strategies: Centralized and decentralized. A compromise between fully centralized and fully decentralized control schemes can be achieved by means of a hierarchical control scheme consisting of three control levels: Primary, secondary, and tertiary. Micro-grids three-level hierarchical control consists of controlling the voltage and frequency, as primary and secondary controls (droop control), while the tertiary control is responsible for economic and optimization operations, including Battery Management System (BMS), control of the Battery Storage System (BSS), as well as managing the import and export energy between the micro-grid and power systems [28,35]. The innovative and integrated energy management and control systems will follow a hierarchical control structure, based on three levels of hierarchy:

- local decentralized control of individual distributed resources;
- regional control of manageable subsets of resources, at micro-grid or VPP (Virtual Power Plant) level; and
- central control of all the system resources, previously aggregated at the second hierarchy level.

The integrated energy management system, for optimizing future aggregators' assets, is coordinated through an advanced ICT data hub platform that allows DSOs (Distribution System Operators) and aggregators to deal with innovative aspects of their distribution network operation and planning with smart energy storage solutions, including V2G integration, power electronics based interface, with bi-directional power control of DER and ESS and integrating IoT into the micro-grids.

Hybrid micro-grids and the integration of DER units introduce several operational challenges that need to be incorporated into the control systems, in order to make certain that the present levels of reliability are not considerably affected, and the potential benefits of DG are fully exploited. The main role of micro-grid control is to assist the active and reactive power generated by the DER components, as well as load demand.

The most relevant provocations in the control of a micro-grid, include bi-directional power flow, stability issues, power balance, uncertainties among DERs coordination, and transition between stand-alone and grid connected operation modes.

Providing technical support to the grid during the transmission of power from generating units to the consumers can be referred to as an ancillary service and involves the adjustments of both the active and reactive power of the flexible resources. For power management in modern micro-grids and the utilization of ESS in different applications, the applied control strategy is the most critical part, in order to have full benefit of all the elements of the power system.

2.4. Integration of RES into Hybrid Micro-Grid Systems

A massive rollout of DERs, together with their real integration in the micro-grid management, is an expected requirement in achieving a real impact on technology. The cost of energy will be a critical factor for driving investments in renewable and alternative energy technologies, and a stable investment climate will be required to exploit new hydrocarbon reserves and develop alternative and renewable energy resources.

A successful hybrid micro-grid deployment depends on a fully integrated system. The integration of RES in hybrid micro-grid systems has become increasingly attractive, mainly because of the minimal operating and maintenance costs. The cost of PV systems and WTs has dramatically declined in the last few years. PV and WT systems also become more efficient due to advancing technologies and communication systems. The hybrid micro-grids combine the benefits of RES with the advantages of conventional power generation, delivering lower, long-term operating costs of ownership.

Three models have been proposed in relation to the energy prosumers that are integrated into the micro-grids: Peer-to-peer, prosumer-to-grid, and prosumer community groups [36]. The first model supports the ability of electricity producers and consumers to directly buy and sell electricity, while the next two methods are close to the centralized grid model concepts and architecture.

Each component of a hybrid micro-grid brings advantages that strengthen the entire system, but energy storage systems are a key enabler, contributing to power stability and energy time-shifting. The deployment of long-term ESS can have an important effect on the optimal operation of micro-grids. In addition to balancing the demand-supply when power shortage or surplus are encountered, ESS can be used to maintain dispatchable DG units operating at their maximum efficiency and can prevent or reduce the use of expensive energy sources during peak hours. More details on ESS will be presented in the next section.

3. The Role of Energy Storage in Micro-Grids

A fundamental partner of the micro-grid improved architecture is the ESS embedded in the architecture, which is described in Section 4. Even if the batteries of the EVs connected to the grid may be used according to the V2G paradigm, an embedded ESS is necessary to provide the grid with at least a minimum amount of storage. The ESS is useful as a tank where the excessive energy arising from the production side can be stored, or by which the lacking energy to sustain a user request can be found. The ESS, thus, acts as a mediator between the energy production and the energy consumption sides, providing several advantages [37]. First of all, the ESS provides a peak shaving function that is useful from both sides. If the peaks in the power demand are sustained by the ESS (e.g., during the fast charge of an EV connected to the grid), the power requirements towards the grid can be relaxed with beneficial effects in both, grid connection costs and technical constraints. On the other hand, ESS will store the peaks in RES energy production due to favorable weather conditions, when the user demand is low or the energy introduction in the main grid is not profitable or even costly. To fully exploit the beneficial effects of the presence of an ESS, a controller that manages the three-way bi-directional energy fluxes is mandatory. The control of the energy fluxes should consider the expected level of energy production (even with possible short and middle term forecasts), the expected demand coming from the user side and the costs, both positive and negative, associated with the power exchange toward the grid. The open research problem is in finding the optimal algorithms with which the energy fluxes are managed and considering the constraints given by the limits in maximum power and maximum energy, which characterize the ESS used in the micro-grid [38–45].

3.1. Energy Storage System Technologies for Micro-Grids

The energy storage technology, that is presently widely acknowledged as the best to use in micro-grids, is based on electrochemical components, in particular rechargeable batteries [12]. Li-ion chemistry is emerging and consolidated as the most used battery technology. The intrinsic advantages of Li-ion batteries with respect to lead-acid stand as the larger values of the energy and power densities with respect to both unit weight (Wh/kg and W/kg) and unit volume (Wh/L and W/L) [46]. The net results are ESSs that are more compact in size and weight, more efficient for the larger number of charge/discharge cycles possible, and even cost-effective in long-term applications. These advantages are only marginally counterbalanced by the larger cost of the battery cells and the poor intrinsic safety of some sub-chemistries of the Li-ion family, which requires an accurate monitoring and management of the battery, in order to keep every cell constituting it in its Safe Operating Area (SOA) [47].

If the Li-ion cell costs are progressively decreasing due to the everyday larger number of applications in which they are used, the development of BMSs, capable of optimally monitoring and managing a Li-ion battery, is still open [48]. Several electronic circuits mainly based on application specific integrated circuits (e.g., battery monitor ICs) are being developed and utilized as BMSs [49–51]. The main and fundamental function of a BMS is to measure most of the battery quantities, and in particular, the voltage and temperature of each cell, the battery current and any other useful parameter, and to operate to maintain every cell of the battery inside the SOA. It is well known that exceeding the SOA limits may lead to dangerous and even catastrophic consequences [52]. Therefore, the BMS must be able to limit or even interrupt the battery current when a potentially dangerous situation is detected. Moreover, the BMS is required to carry out other more advanced functions, such as the charge balancing between the cells to compensate possible mismatches, the estimate of the residual charge available in the battery, the estimate of the state of health of the battery cells, and to perform all the diagnostic functions useful to reveal the battery status and prolong battery life [53]. The battery temperature management is another task in charge of a BMS, usually carried out by activating fans for the battery cooling or even heaters, when the battery operates at very low temperatures and needs to be pre-heated before being fully operational. BMSs are sometimes also asked to dynamically manage the battery architecture [54,55], by controlling the parallelization of battery strings when the battery architecture is modular, and the battery capacity in ampere-hour can be selected and adapted to the application requirements by choosing the number of strings to be connected in parallel. Even part of the battery module series, that increase the battery voltage, can be excluded by the BMS with a bypass switch, in order to allow the module substitution or maintenance without interrupting the energy storage service [56]. Finally, the BMS provides the communication functions required to insert the ESS in the micro-grid network, by which, the grid management functions and the policies regarding the energy flux management, are carried out. In conclusion, the BMS is an electronic system that provides very powerful and fundamental functions. The BMS realization is based on electronic platforms that must be sufficiently powerful but, at the same time, compatible in size and cost affordable for the considered application. Microcontroller-based and even-FPGA (Field Programmable Gate Array) based platforms, when complex algorithms have to be executed, are the most useful solutions as BMS realizations [57]. Figure 4 shows the main functions performed by a BMS, grouped in the three fields of Safety and Lifetime Extension, Battery State Estimation and System Integration.

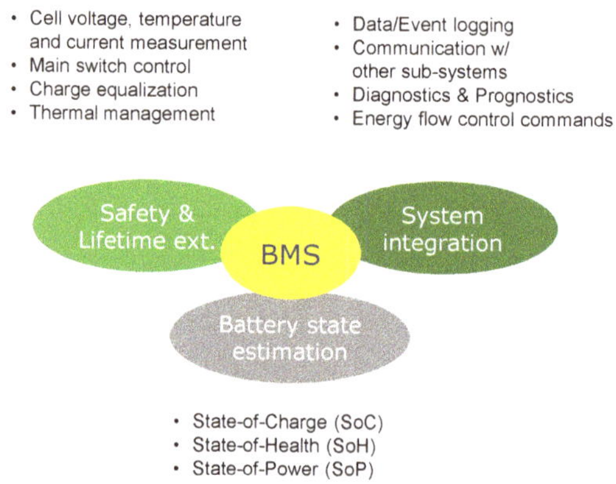

Figure 4. Main functions performed by a Battery Management System (BMS).

3.2. Non Lithium Battery Technologies for Micro-Grids

Besides lead acid batteries, other electrochemical technologies are being used as storage systems in stationary applications. Sodium-sulfur and sodium-metal halide batteries (also known as ZEBRA batteries [58]) have been applied in installations beside utility factories and PV fields [59]. These technologies may also be useful in smaller stationary applications, such as in micro-grids, because of the intrinsic safety of the technology, which does not require the accurate and safe management of the cells. The technologies are based on the conduction of sodium ions through a beta-alumina electrolyte, and thus, employ a low-cost material as sodium that is abundant in nature. The only disadvantage is the necessity to operate the battery at an inner temperature of at least 250 °C to allow the liquefaction of the sodium and the ion conduction. Therefore, heaters consuming useful energy are required to maintain the operation of the battery. An interesting feature of sodium-metal halide batteries is that when a cell fails, it results in a short circuit that maintains the continuity and the operation of a series connected string, even if with a slightly lesser overall voltage. Unfortunately, the technology is promising but has not reached full development yet, in order to deeply penetrate the market. Further improvement in cell reliability and BMSs are needed to fully exploit the possible technology advantages [60].

3.3. Advanced Battery Management Systems

As pointed out in Section 3.1, the BMS plays a fundamental role in every energy storage system [61], particularly those based on Li-ion technology. Besides the basic monitoring and management functions mentioned above, there are advanced functions that updated BMS is required to perform. First of all, the BMS architecture must be designed to fit to different configurations of the battery, that usually consists of cells, modules (usually series connected), and strings (usually parallel connected). The BMS functions may be spread among the various layers of the BMS architecture [62]. Typical architectures span from a single master BMS that controls the entire battery, to hierarchical distributed ones, by which the BMS hardware is distributed over the battery modules and even down the individual battery cells [63,64]. Figure 5 shows a typical BMS hierarchical architecture that matches the parallel/series structure of the battery cells [49,65].

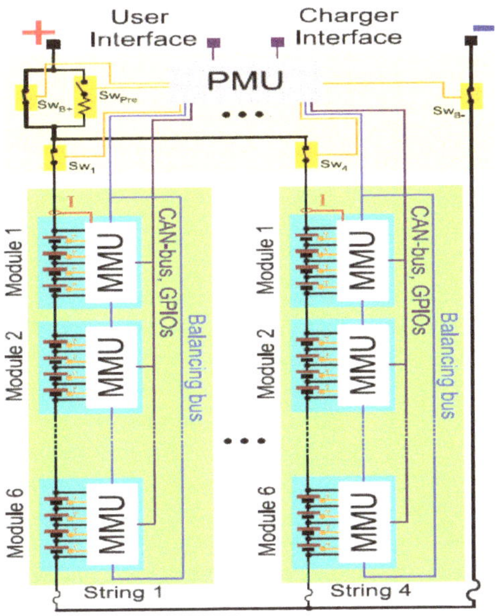

Figure 5. Hierarchical architecture of a BMS matching the battery structure consisting of four paralleled strings of 6 series connected modules each. The Pack Management Unit (PMU) is the master BMS, connected to the lower layer slave BMSs (MMUs) [49].

The State of Charge (SoC) of a battery is a fundamental quantity as it indicates the residual charge remaining in the battery, and thus, gives indications on the residual support provided by the ESS to the application. Moreover, the SoC estimate must be carried out on every cell of the battery, as the battery cells may be mismatched, and that mismatch may increase with ageing. Unfortunately, SoC is a quantity not directly measurable, but it must be calculated from other quantities, e.g., by integrating, in time, the current flowing in the battery. As in every non-ideal integration process, offsets or inaccuracies in the current sensor readings rapidly lead to a calculation of the integral that is affected by increasing and unacceptable errors. Many alternative approaches to the so-called Coulomb counting technique have been proposed [66,67]. Most of them are based on a model of the battery that is subjected to the same load and operating conditions of the real battery. The status of the battery, estimated from the model calculation, resembles the status of the real battery. Therefore, the SoC and other battery parameters can be estimated by the model. The most popular models are electrical models. They provide acceptable accuracy, together with affordable computation complexity. The battery is represented by an electromotive force (the so-called Open Circuit Voltage) that varies as a function of SoC, a series resistance, that models the internal losses, as well as a series of some resistor-capacitor parallel groups, which model the relaxation phenomena in the battery. The BMS is, thus, required to solve the model-based circuit to obtain an accurate estimate of the SoC. A further effect, that makes the SoC estimate tougher, is the variation of the model parameters. In fact, the values of the electrical components, which make up the battery model, vary in time, according to the state of charge, temperature, the environmental conditions, and the health status of the battery. A further challenge for an advanced BMS is, thus, to identify and track the parameter changes, only by observing the current, voltages, and temperatures of the battery cells.

Several state estimation and parameter identification techniques have been applied in the literature, starting from the moving window least square methods, to filtering, such as Extended Kalman Filters and Particle Filters. In any case the choice of the estimation technique must be traded off between the computation complexity required and the accuracy needed.

An even tougher task for a BMS is the evaluation of the health status of a battery [68,69]. The battery ageing is a complex process that depends on the utilization of the battery in term of operating temperatures, load profiles, and the number of charge/discharge cycles. It typically results in the progressive fading of the maximum capacity of the battery and the increase of the series resistance, which determine the increased internal losses and reduced power capability. The knowledge of the battery health status, together with the expected operational profile, may lead the BMS to estimate the Residual Useful Life of the battery.

3.4. Battery Charging Process

Battery cell manufacturers usually provide the user with constraints about the appropriate recharging process of the battery. The suggested charging profile is the Constant-Current Constant-Voltage (CC-CV) one [70]. It consists of a first phase in which the battery is charged with a constant current. The current value depends on the cell technology, and it is in the order of the battery capacity C expressed in ampere. Fast charging usually exceeds 1 C, whereas slow charging spans from 0.1 C to 0.5 C. The faster the charging, the higher the internal cell temperature and thus the possible stress that may determine ageing. The second phase starts when the battery voltage reaches the float value. The charger keeps the voltage constant since then, to gradually fill the battery up the maximum charge level. The process ends when the decreasing current reaches a threshold, usually a rather low fraction of the battery capacity expressed in ampere. The battery inserted in the multi-source/sink micro-grid scenario, depicted in Figure 6, cannot fully adhere to the above described charging process. Instead, as the charging and discharging phases are defined according to the overall management of the energy fluxes, the BMS should verify the manufacturer constraints in terms of current, temperature, and individual cell voltages. Should it happen, the bi-directional DC/DC converter, that connects the battery to the power bus (see detailed circuit schematic in Figure 7), must be driven in such a way as to reduce the charging current and to withstand the manufacturer's limits. As the battery is not subjected to full discharge/charge cycles, the accurate knowledge of the battery SoC is mandatory to fully exploit the established energy flux policy mentioned in Section 2.3.

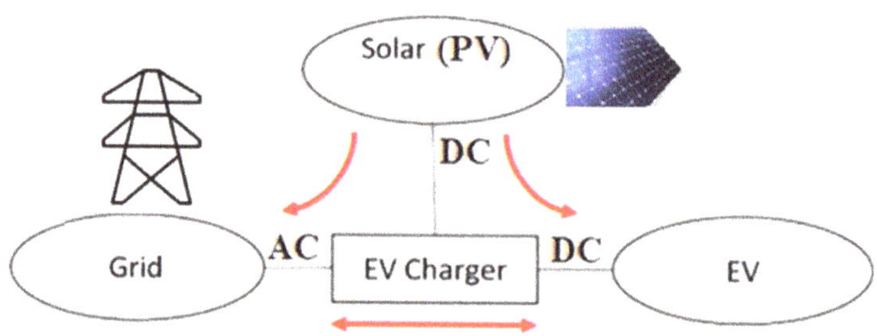

Figure 6. New hybrid micro-grid with full connectivity among RES node (DC), EV node (DC) and main grid node (AC) and relevant bi-directional flows. EV: Electric Vehicle.

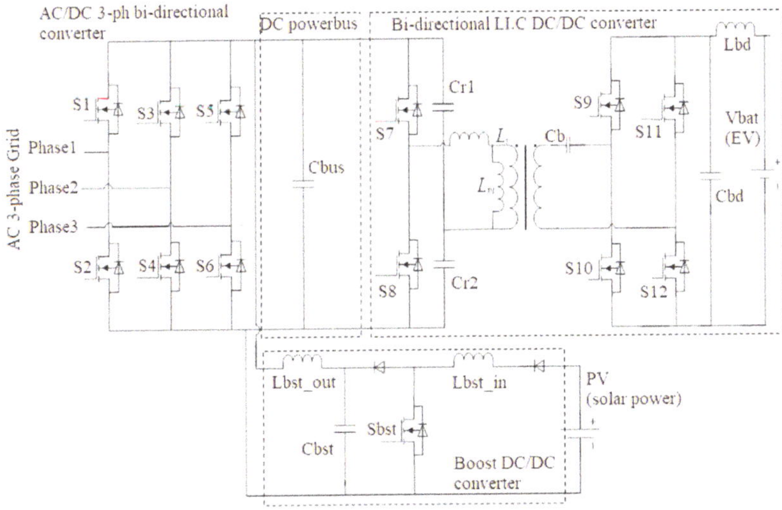

Figure 7. Circuit schematic of the bi-directional converter (1 AC grid port and 1 DC EV battery bidirectional ports; 1 DC unidirectional solar power port).

4. Innovative Micro-Grid with Bi-Directional Flows for RES and EV Charging

As anticipated above, a new micro-grid architecture is described in this Section (see Figure 6), which exploits bi-directional EV charger plus RES to overcome the flexibility limit of the architecture, as shown in Figure 1. Section 3 has shown why Li-ion batteries are being widely used for the high power and energy densities. Thus, they serve as energy storage unit connecting to the smart grid and to a RES such as PV panels in our focused application. This leads to a hub connection discussion: The current industry devices lack an efficient integrated connection system, since PV panels are not directly connected to the EV charger. Typically, they are first connected to the grid via a unidirectional DC/AC solar inverter, then a traditional unidirectional EV charger is used to connect the grid with the EV implementing an AC/DC conversion. This means that energy flows over two paths in a conventional solar-powered EV charging system, resulting in extra power losses and higher system cost. Instead, a novel bidirectional EV charger system is proposed in Figure 6 to build a direct connection between PV panels and the EV, and to create a V2G path. Therefore, reduced power loss and lower system cost features are achieved from a highly integrated power electronics system with high efficiency and high-power density.

The idea of the system scheme in Figure 6, as further detailed in the circuit scheme of Figure 7, is having a central DC power bus, plus a bidirectional flow between the AC grid and the power DC bus thanks to a 3-phase bidirectional converter, and a bidirectional flow between the DC battery on-board the EV and the DC power bus thanks to a bidirectional DC/DC converter. A unidirectional flow is still foreseen from the PV solar panel sub-system towards the DC power bus, using a boost DC/DC converter to adapt the output solar PV level to the DC power bus level.

The DC power bus voltage is typically in the range from 250 V to 600 V, e.g. it has been sized at 400 V in [3] in case of a 10 kW bidirectional EV charger and at 450 V in case of a 1.65 kW bidirectional EV charger in [71]. In [72–74], as special optimization case, a bidirectional EV charger is proposed where the value of the DC power bus can be sized from 500V to 840V. The AC grid is typically a 3-phase 380Vac one.

In the detailed circuit schematic of the EV bidirectional charger in Figure 7 the bidirectional 3-phase AC/DC converter and the boost DC/DC converter follow classic circuit solutions. For the isolated bi-directional DC/DC converter in Figure 7, instead of using a dual-active bridge topology as in [71–74] (see Figure 8B), a half-bridge series resonant LLC topology is proposed (that of Figure 8A). This approach allows reducing the number of active switches to be used and hence it makes more convenient the adoption of SiC power Mosfets (e.g., Cree C2M0040120D adopted in [3]) instead of classic Si power Mosfets (e.g. Infineon IPW60R045CP adopted for the primary stage in [71]). Indeed, SiC power Mosfets are more expensive than Si power Mosfets: a market analysis on stocks from 100 to 1000 devices for the power mosfets in Table 1 has shown that the selling price for each SiC power device is 1.7 to 3.5 times higher than that of Si power device. Hence, a circuit solution like that in Figure 8A that minimizes the use of active devices make more convenient the adoption of SiC power Mosfets.

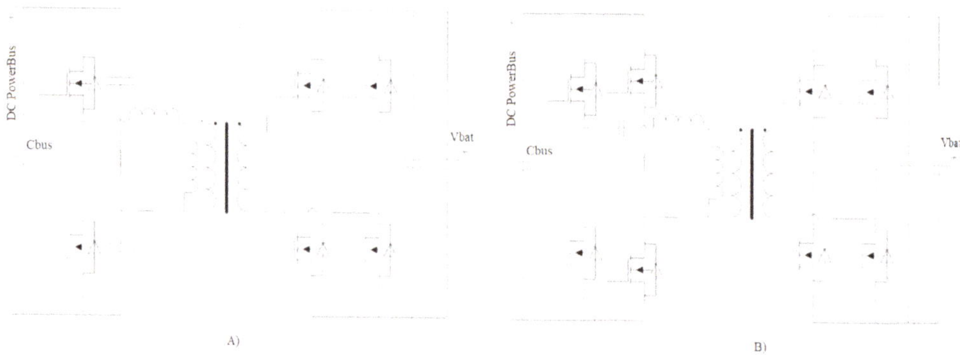

Figure 8. Bi-directional DC/DC converter: (**A**) half-bridge at the primary stage; (**B**) dual-active bridge topology.

Table 1. Power Mosfets performance comparison (same package and current), SiC in [3] vs. Si in [71].

Power Mosfets	Package	I_{Dmax}	V_{DSmax}	R_{DSon}	Q_G	T_{dON}	T_{dOFF}	T_{rr}
SiC C2M0040120D	TO-247	60 A	1200 V	40 mΩ	115 nC	15 ns	26 ns	54 ns
Si IPW60R045CP	TO-247	60 A	60 V	45 mΩ	150 nC	30 ns	100 ns	600 ns

As shown in Table 1, for the same package (T0-247) and hence the same occupied area on a printed circuit board, and the same rated current I_{Dmax}, SiC power Mosfets double the sustained voltage vs. Si power Mosfets (1200 V vs. 600 V), improve the typical figure of merit R_{DSon} x Q_G [75] by reducing both the R_{DSon} and the total gate charge (Q_G), and remarkably reduce ton/off-delay time (t_{dON}/t_{dOFF}) and reverse recovery time (t_{rr}) of the integrated diode.

From a switching frequency point of view, the solutions proposed in [3,71] for bidirectional DC/DC conversion are in the order of 100 kHz, more in detail up to 140 kHz in [71] and up to 150 kHz in [3]. As discussed in [72–74], by using wide bandgap power devices like SiC and GaN the typical frequencies of power Si devices of about 100 kHz can be increased above 300 kHz. For example, a 500 kHz switching frequency is used for a 6.6 kW CLLC DC/DC resonant converter in [72–74] where a bi-directional EV charger is proposed by cascading a first AC/DC stage (switching at 300 kHz) followed by the 500 kHz CLLC DC/DC converter stage. The higher the frequency, the lower the size and weight of the inductive passive devices (inductors, transformers).

As far as the selection of the active devices is concerned the work in [72–74] proposes a mix of 650 V GaN and 1.2 kV SiC power switch devices. Particularly, since in [72–74] the power DC Bus of the EV charger is sized for values ranging from 500 V_{DC} to 840 V_{DC} (when the battery on board the EV is 250 V_{DC} to 420 V_{DC}) it is necessary to change the 650V GaN devices with 1.2 kV SiC Mosfets in the AC/DC stage and in the primary side of the DC/DC stage of the bidirectional EV charger. 650V GaN devices are used in [72–74] only for the secondary stage of the bidirectional DC/DC converter. A full

650 V GaN devices solution is discussed in [74] but limited to applications where the DC power bus is 450 V at maximum.

Summarizing, compared with Si devices, the absence of reverse recovery charge in wide-bandgap devices (GaN and SiC) enables bidirectional operation with simplified converter structures. Moreover, wide-bandgap devices have better figure of merit [75] than Si devices and hence for a given ON-resistance and breakdown voltage, wide-bandgap devices require smaller die size, which translates into smaller gate charge and junction capacitance. These characteristics allow increasing switching frequency (above the 100 kHz typical of converters with power SI devices) and reducing the switching loss. From reported results [3,72–74] of recent bi-directional EV chargers, 650 V GaN devices may allow a higher switching frequency than 1.2 kV SiC devices, but the latter become mandatory to manage in a reliable and efficient manner DC power bus of 500 V_{DC} or above.

With the larger battery capability in EVs, power flow between the EV and the grid is exploited in a bi-directional fashion. It is especially useful in applications using renewables such as solar generation as in [3]. Its flow chart is viewed in Figure 6, in contrast to the traditional way shown in Figure 1. In the bi-directional power flow interaction, the solar power can flow either from the PV panel to the grid or to the EV, and the battery charging power can flow from grid to vehicle (G2V) or vice versa (V2G). Two additional power flows are provided and managed in Figure 6 if compared to the unidirectional mode in Figure 1. Moreover, a further partner of the new architecture can also be the ESS, discussed in Section 3, located inside the residential installation that may or may not be included in the above described scenario.

The proposed approach in Figure 6 has several main advantages. First, there is an increase in system flexibility. V2G power flow implies the power can be drawn from the battery serving as the electric source in [4], which has significant benefits in the future applications such as powering house during the electricity shutdown. Moreover, the system becomes quite simpler after applying the bi-directional power flow method, and its benefits can be summarized in these two points: (i) Direct access to solar generation and EV battery; (ii) efficiency maximization.

Concerning the direct access to solar generation and EV battery, thanks to the two additional power flow path—solar power to EV charging and EV back-feed to the grid-(see Figure 6 compared to Figure 1), the system becomes more valuable in the energy usage and transfer. The system integrates the functionality of solar inverter and battery charging into one system, by providing direct access between each power element, which significantly reduces the system size.

Concerning efficiency maximization, the overall system efficiency can be improved of about 4% by comparing the bi-directional power flow method with the traditional unidirectional power flow method. From the perspective of fairness, their measured efficiency data are based on same metrics of power electronics system experiment: EV charger data are taken from [3], and solar inverter data are taken from [5]. If for example the power converters (solar inverter and unidirectional converter in Figure 1, and bi-directional converter in Figure 6) have an efficiency of 96%, the scheme in Figure 1 will result in an overall efficiency of 92% less than the efficiency of 96% achieved with the scheme in Figure 6.

The 4% improvement is a substantial saving on the economics of electricity usage, besides the materials cost saving from one compact integrated system, compared to two bulky systems separated systems. The only hardware requirement to implement the circuits and systems proposed in Figures 6 and 7 is the use of bi-directional capable semiconductor switches in [3], based on the today available SiC technology. Since SiC is becoming a mature technology, the proposed approach adds trivial circuit complexity as shown in the circuit in Figure 7, which has been implemented and characterized in [3,21–23]. The bi-directional converter in Figure 7 has been tested with battery packs of 400 V and charging/discharging power levels in the range from 1 to 10 kW.

It is worth noting that the scheme in Figure 6, and hence the circuit in Figure 7, can further be improved in terms of flexibility since it is possible to exploit as ESS, not only where available in the EV side, but also that available at RES side. Indeed, new solar energy sources are available in the market [2], where a high-voltage Li-ion battery energy storage is integrated with capacity scalable from 4 kWh up to 12 kWh. The energy storage in [2] can sustain DC voltage inputs from 170 V to 575 V at the battery port side, where each basic battery module of 4 kWh can be charged/discharged at maximum charge power of 1.6 kW, and maximum discharge power of 2 kW. By parallelizing the modules, we obtained an overall energy storage up to 12 kWh with maximum charge power of 4.8 kW and maximum discharge power of 6 kW. In this way, new additional modes with respect to Figure 6 can also be enabled, with power flowing also towards, and not only from, the solar power node. Moreover, the new products appearing on the market for energy production, conversion and storage are equipped with advanced connectivity and diagnostic features [76,77], helping the integration phase in micro/smart grids.

5. Conclusions

The cost-reducing techniques of RES components used in micro-grids are mainly PV systems and WTs, and these provide new possibilities for clean, reliable, and affordable systems in micro-grid configurations. New hybrid micro-grid systems can quickly and cost-effectively be implemented in remote locations, thus, opening new opportunities to investigate the new concepts as alternatives to conventional power systems.

ESS components, particularly batteries based on Li-ion or sodium technologies, are identified as a key technology for the integration of intermittent RES. This, in turn, introduces major challenges to the control system for the appropriate management of this resource.

The integration of an energy management system in a future micro-grid, for optimizing future aggregators' assets, should be coordinated through an advanced ICT data hub platform that allows DSOs and aggregators to deal with innovative aspects of their distribution network operation and planning with smart energy storage solutions. This would include V2G integration, power electronics-based interface with bi-directional power control of DER and ESS, and integrating IoT (Internet of Things) into the Smart Grids.

With reference to a new and improved energy grid, this paper first reviewed trends in RES, power converters, and control strategies, and analyzed trends in battery energy storage and the relevant issues in battery charging and BMS. An alternative micro-grid architecture was then proposed to overcome the lack of flexibility of the classic energy grid. By mixing DC and AC sources, the hybrid micro-grid shows an alternative architecture where the use of bi-directional EV chargers may allow the creation of a micro-grid directly interconnecting all the nodes with bi-directional energy flows: Main grid node, RES node, energy storage nodes, both on-board the vehicle and inside the micro-grid structure. The proposed model is also sustained by the new products emerging in the market, such as new solar inverters that include are appearing [2], where a local energy storage for the RES and hence can enable bidirectional power flows from/towards the RES.

Author Contributions: Research: S.S., R.S., and L.M.-P. Writing: S.S., R.S., and L.M.-P.

Funding: This work was partially supported by Tuscany Region under "Progetto SUMA" and partially supported by CrossLab project, University of Pisa, funded by MIUR "Department of Excellence" program.

Acknowledgments: Discussions with K.J., L.C., W.X. from MIT, US under the MIT-UNIPI seed fund program are gratefully acknowledged.

Conflicts of Interest: The authors declare no conflict of interest.

References

1. Mihet-Popa, L.; Saponara, S. Toward Green Vehicles Digitalization for the Next Generation of Connected and Electrified Transport Systems. *Energies* **2018**, *11*, 3124. [CrossRef]
2. Available online: https://new.abb.com/power-converters-inverters/solar/photovoltaic-energy-storage/react-2 (accessed on 16 November 2019).

3. Wang, X.; Jiang, C.; Lei, B.; Teng, H.; Bai, H.K.; Kirtley, J.L. Power-loss analysis and efficiency maximization of a silicon-carbide MOSFET-based three-phase 10-kW bidirectional EV charger using variable-DC-bus control. *IEEE J. Emerg. Sel. Top. Power Electron.* **2016**, *4*, 880–892. [CrossRef]
4. Jain, P.; Jain, T. Impacts of G2V and V2G power on electricity demand profile. In Proceedings of the 2014 IEEE International Electric Vehicle Conference (IEVC), Florence, Italy, 17–19 December 2014.
5. Wall, S.; Ruan, R.; Wang, C.; Xie, J. Evaluation of three-phase solar inverters using SiC devices. In Proceedings of the IEEE EPE'16 ECCE Europe, Karlsruhe, Germany, 5–9 September 2016.
6. Olivares, D.E.; Mehrizi-Sani, A.; Etemadi, A.H.; Cañizares, C.A.; Iravani, R.; Kazerani, M.; Hajimiragha, A.H.; Gomis-Bellmunt, O.; Saeedifard, M.; Palma-Behnke, R.; et al. Trends in Micro-grid Control. *IEEE Trans. Smart Grid* **2014**, *5*, 1905–1919. [CrossRef]
7. Hirsch, A.; Parag, Y.; Guerrero, J. Micro-grids: A review of technologies, key drivers, and outstanding issues. *Renew. Sustain. Energy Rev.* **2018**, *90*, 402–411. [CrossRef]
8. Sánchez-Miralles, M.M.F.; Rivier, A.M. A literature review of micro-grids: A functional layer based classification. *Renew. Sustain. Energy* **2016**, *62*, 1133–1153.
9. Vadi, S.; Padmanaban, S.; Bayindir, R.; Blaabjerg, F.; Mihet-Popa, L. A Review on Optimization and Control Methods Used to Provide Transient Stability in Micro-grids. *Energies* **2019**, *12*, 3582. [CrossRef]
10. Ganesan, S.; Padmanaban, S.; Varadarajan, R.; Subramaniam, U.; Mihet-Popa, L. Study and Analysis of an Intelligent Micro-grid Energy Management Solution with Distributed Energy Sources. *Energies* **2017**, *10*, 1419. [CrossRef]
11. Ton, D.; Reilly, J. Micro-grid Controller Initiatives. *IEEE Power Energy Mag.* **2017**, *15*, 24–31. [CrossRef]
12. Xavier Saury, F.; Tomlinson, C. *Hybrid Micro-Grids: The Time Is Now*; CATERPILLAR-LEXE0904; Peoria, IL, USA, 2016; Available online: https://www.cat.com/en_US/by-industry/electric-power-generation/Articles/White-papers/white-paper-hybrid-microgrids-the-time-is-now.html (accessed on 28 September 2019).
13. Xiao, Y.; Ren, C.; Han, X.; Wang, P. A generalized and mode-adaptive approach to the power flow analysis of the isolated hybrid AC/DC micro-grids. *Energies* **2019**, *12*, 2253. [CrossRef]
14. Han, Y.; Li, H.; Shen, P.; Coelho, E.A.A.; Guerrero, J.M. Review of active and reactive power sharing strategies in hierarchical controlled micro-grids. *IEEE Trans. Power Electron.* **2017**, *32*, 2427–2451. [CrossRef]
15. Unamunon, E.; Barrena, J.A. Hybrid AC/DC microgrids-Part II: Review and classification of control strategies. *Renew. Sustain. Energy Rev.* **2012**, *52*, 1123–1134. [CrossRef]
16. Mani Venkata, S.S.; Shahidehpour, M. Microgrid Controllers: The Brain, Heart, & Soul of Microgrid Automation. *IEEE Power Energy Mag.* **2017**, *15*, 16–22.
17. Colak, I.; Kabalci, E.; Fulli, G.; Lazarou, S. A survey on the contributions of power electronics to smart grid systems. *Renew. Sustain. Energy Rev.* **2015**, *47*, 562–579. [CrossRef]
18. Mohammed, A.; Refaat, S.S.; Bayhan, S.; Abu-Rub, H. AC Microgrid Control and Management Strategies-Evaluation and review. *IEEE Power Electron. Mag.* **2019**, *6*, 18–31. [CrossRef]
19. Shen, X.; Tan, D.; Shuai, Z.; Luo, A. Control techniques for bidirectional interlinking converters in Hybrid Microgrids. *IEEE Power Electron. Mag.* **2019**, *6*, 39–47. [CrossRef]
20. Bhalla, A. Electromagnetic interference mitigation in wide-bandgap power conversion. *IEEE Power Electron. Mag.* **2019**, *6*, 32–35. [CrossRef]
21. Adan, A.O.; Tanaka, D.; Burgyan, L.; Kakizaki, Y. The current status and trends of 1200-V commercial silicon-carbide MOSFETs. *IEEE Power Electron. Mag.* **2019**, *6*, 36–47. [CrossRef]
22. Bindra, A. Wide-bandgap power devices: Adoption gathers momentum. *IEEE Power Electron. Mag.* **2018**, 22–27. [CrossRef]
23. Mookken, J. SiC MOSFETs enable high frequency in high power conversion systems. *Bodo's Power Syst.* **2016**, 28–32. Available online: https://www.richardsonrfpd.com/docs/rfpd/SiC_MOSFET_Article.pdf (accessed on 16 November 2019).
24. International Energy Agency-IEA. Renewables 2018. Market Analysis and Forecast from 2018 to 2023, October 2018; ISBN 978-92-64-36998-6. Available online: https://www.iea.org/renewables2018/ (accessed on 16 November 2019).
25. International Energy Agency-IEA. Renewables 2019. Market Analysis and Forecast from 2019 to 2024, October 2019; ISBN 978-92-64-30684-4. Available online: https://www.iea.org/renewables2019/ (accessed on 16 November 2019).

26. Shivarama Krishna, K.; Sathish Kumar, K. A review on hybrid renewable energy systems. *Renew. Sustain. Energy Rev.* **2015**, *52*, 907–916. [CrossRef]
27. Rocabert, J.; Luna, A.; Blaabjerg, F.; Rodríguez, P. Control of power converters in AC microgrids. *IEEE Trans. Power Electron.* **2012**, *27*, 4734–4749. [CrossRef]
28. Zang, M.; Li, Y.; Peng, Y.; Li, W.; Liu, F. A control architecture to coordinate DG inverters a series-LC-filter hybrid active filter for power quality improvement in Micro-Grid. In Proceedings of the 2017 IEEE International Conference on Environment and Electrical Engineering and 2017 IEEE Industrial and Commercial Power Systems Europe (EEEIC/I & CPS Europe), Milan, Italy, 13 July 2017.
29. Cirrincione, M.; Pucci, M.; Vitale, G. A single-phase DG generation unit with shunt active power filter capability by adaptive neural filtering. *IEEE Trans. Ind. Electron.* **2008**, *5*, 2093–2110. [CrossRef]
30. Yu, Y.; Guerrero, J.M.; Sun, L.; Li, H. Modeling and stability analysis of CCIs- and VCIs- Based Hybrid Microgrids operating in Grid-Connected Modes. In Proceedings of the 2018 International Conference on Smart Energy Systems and Technologies (SEST), Sevilla, Spain, 10–12 September 2018.
31. Luo, F.; Loo, K.H.; Lai, Y.M. A Hybrid AC/DC Microgrid Control Scheme with Voltage-Source Inverter-Controlled Interlinking Converters. In Proceedings of the 18th European Conference on Power Electronics and Applications, EPE 2016 ECCE Europe, Karlsruhe, Germany, 5–9 September 2016.
32. Peyghami, S.; Mokhtari, H.; Blaabjerg, F. Autonomous operation of a Hybrid AC/DC Microgrid with multiple interlinking converters. *IEEE Trans. Smart Grid* **2018**, *19*, 6480–6488. [CrossRef]
33. Liu, Q.; Caldognetto, T.; Buso, S. Flexible control of interlinking converters for DC Microgrids coupled to smart AC power systems. *IEEE Trans. Ind. Electron.* **2019**, *66*, 3477–3485. [CrossRef]
34. Guerrero, J.M.; Loh, P.C.; Lee, T.L.; Chandorkar, M. Advanced control architectures for intelligent microgrids—Part II: Power quality, energy storage, and AC/DC microgrids. *IEEE Trans. Ind. Electron.* **2013**, *60*, 1263–1270. [CrossRef]
35. Sadeghkhani, I.; Golshan, M.E.H.; Mehrizi-Sani, A.; Guerrero, J. Low voltage ride-through of a droop-based three-phase four-wire grid-connected microgrid. *IET Gener. Transm. Distrib.* **2018**, *12*, 1906–1914. [CrossRef]
36. Mihaylov, M.; Razo-Zapata, I.; Rădulescu, R.; Nowé, A. Boosting the Renewable Energy Economy with NRGcoin. In Proceedings of the 4th International Conference on ICT for Sustainability (ICT4S), Amsterdam, The Netherlands, 29 August–1 September 2016.
37. Beaudin, M.; Zareipour, H.; Schellenberg, A.; Rosehart, W. Energy Storage for Mitigating the Variability of Renewable Electricity Sources. *Energy Storage Smart Grids* **2010**, *14*, 302–314.
38. Zhao, T.; Ding, Z. Cooperative Optimal Control of Battery Energy Storage System under Wind Uncertainties in a Micro-grid. *IEEE Trans. Power Syst.* **2018**, *33*, 2292–2300. [CrossRef]
39. Xie, H.; Teng, X.; Xu, Y.; Wang, Y. Optimal Energy Storage Sizing for Networked Micro-grids Considering Reliability and Resilience. *IEEE Access* **2019**, *7*, 86336–86348. [CrossRef]
40. Khodabakhsh, R.; Sirouspour, S. Optimal Control of Energy Storage in a Micro-grid by Minimizing Conditional Value-at-Risk. *IEEE Trans. Sustain. Energy* **2016**, *7*, 1264–1273. [CrossRef]
41. Ju, C.; Wang, P.; Goel, L.; Xu, Y. A Two-Layer Energy Management System for Micro-grids With Hybrid Energy Storage Considering Degradation Costs. *IEEE Trans. Smart Grid* **2018**, *9*, 6047–6057. [CrossRef]
42. Qi, X.; Bai, Y.; Luo, H.; Zhang, Y.; Zhou, G.; Wei, Z. Novel Distributed Optimal Control of Battery Energy Storage System in an Islanded Microgrid with Fast Frequency Recovery. *Energies* **2018**, *11*, 1955. [CrossRef]
43. García Vera, Y.E.; Dufo-López, R.; Bernal-Agustín, J.L. Energy Management in Microgrids with Renewable Energy Sources: A Literature Review. *Appl. Sci.* **2019**, *9*, 3854. [CrossRef]
44. Ross, M.; Abbey, C.; Bouffard, F.; Joós, G. Micro-grid Economic Dispatch with Energy Storage Systems. *IEEE Trans. Smart Grid* **2018**, *9*, 3039–3047. [CrossRef]
45. Faisal, M.; Hannan, M.A.; Ker, P.J.; Hussain, A.; Mansor, M.B.; Blaabjerg, F. Review of Energy Storage System Technologies in Micro-grid Applications: Issues and Challenges. *IEEE Access* **2018**, *6*, 35143–35164. [CrossRef]
46. Nadeem, F.; Hussain, S.S.; Tiwari, P.K.; Goswami, A.K.; Ustun, T.S. Comparative Review of Energy Storage Systems, Their Roles, and Impacts on Future Power Systems. *IEEE Access* **2019**, *7*, 4555–4585. [CrossRef]
47. Rahimi-Eichi, H.; Ojha, U.; Baronti, F.; Chow, M.-Y. Battery Management System: An Overview of Its Application in the Smart Grid and Electric Vehicles. *IEEE Ind. Electron. Mag.* **2013**, *7*, 4–16. [CrossRef]
48. Brandl, M.; Gall, H.; Wenger, M.; Lorentz, V.; Giegerich, M.; Baronti, F.; Fantechi, G.; Fanucci, L.; Roncella, R.; Saletti, R.; et al. Batteries and battery management systems for electric vehicles. In Proceedings of

the 2012 Design, Automation & Test. in Europe Conference & Exhibition (DATE), Dresden, Germany, 12–16 March 2012.
49. Analog Device 12 Channel Multicell Battery Monitor with Addressable Interface. Available online: https://www.analog.com/media/en/technical-documentation/data-sheets/680412fc.pdf (accessed on 16 November 2019).
50. Maxim Integrated MAX14920 High-Accuracy 12-/16-Cell Measurement Analog Front Ends. Available online: https://datasheets.maximintegrated.com/en/ds/MAX14920-MAX14921.pdf (accessed on 16 November 2019).
51. Texas Instruments bq76PL455A 16-Cell Industrial Integrated Battery Monitor with Passive Cell Balancing. Available online: http://www.ti.com/lit/ds/symlink/bq76pl455a.pdf (accessed on 16 November 2019).
52. Lahiri, A.; Shah, N.; Dales, C. Building a safer, denser lithium-ion battery. *IEEE Spectr.* **2018**, *55*, 34–39. [CrossRef]
53. Omariba, Z.B.; Zhang, L.; Sun, D. Review of Battery Cell Balancing Methodologies for Optimizing Battery Pack Performance in Electric Vehicles. *IEEE Access* **2019**, *7*, 129335–129352. [CrossRef]
54. Ci, S.; Lin, N.; Wu, D. Reconfigurable Battery Techniques and Systems: A Survey. *IEEE Access* **2016**, *4*, 1175–1189. [CrossRef]
55. Lin, N.; Ci, S.; Wu, D.; Guo, H. An Optimization Framework for Dynamically Reconfigurable Battery Systems. *IEEE Trans. Energy Convers.* **2018**, *33*, 1669–1676. [CrossRef]
56. Baronti, F.; Fantechi, G.; Roncella, R.; Saletti, R. Design of a module switch for battery pack reconfiguration in high-power applications. In Proceedings of the 2012 IEEE International Symposium on Industrial Electronics (ISIE), Hangzhou, China, 28–31 May 2012; pp. 1330–1335. [CrossRef]
57. Morello, R.; Baronti, F.; Tian, X.; Chau, T.; Di Rienzo, R.; Roncella, R.; Saletti, R. Hardware-in-the-loop simulation of FPGA-based state estimators for electric vehicle batteries. In Proceedings of the IEEE International Symposium on Industrial Electronics, Santa Clara, CA, USA, 8–10 June 2016. [CrossRef]
58. Sudworth, J.L. Zebra batteries. *J. Power Sources* **1994**, *51*, 105–114. [CrossRef]
59. Moseley, P.T.; Rand, D.A. Chapter 15-High-Temperature Sodium Batteries for Energy Storage. In *Electrochemical Energy Storage for Renewable Sources and Grid Balancing*, 1st ed.; Moseley, P.T., Garche, J., Eds.; Elsevier: London, UK, 2015; pp. 253–268. ISBN 978-044-462-616-5.
60. Li, G.; Lu, X.; Kim, J.Y.; Lemmon, J.P.; Sprenkle, V.L. Improved cycling behavior of ZEBRA battery operated at intermediate temperature of 175 °C. *J. Power Sources* **2014**, *249*, 414–417. [CrossRef]
61. Lawder, M.T.; Suthar, B.; Northrop, P.W.; De, S.; Hoff, C.M.; Leitermann, O.; Subramanian, V.R. Battery Energy Storage System (BESS) and Battery Management System (BMS) for Grid-Scale Applications. *Proc. IEEE* **2014**, *102*, 1014–1030. [CrossRef]
62. Kim, H.; Shin, K.G. DESA: Dependable, Efficient, Scalable Architecture for Management of Large-Scale Batteries. *IEEE Trans. Ind. Inform.* **2012**, *8*, 406–417. [CrossRef]
63. Steinhorst, S.; Lukasiewycz, M.; Narayanaswamy, S.; Kauer, M.; Chakraborty, S. Smart Cells for Embedded Battery Management. In Proceedings of the 2014 IEEE International Conference on Cyber-Physical Systems, Networks, and Applications, Hong Kong, China, 25–26 August 2014.
64. Baronti, F.; Fantechi, G.; Roncella, R.; Saletti, R. Intelligent cell gauge for a hierarchical battery management system. In Proceedings of the 2012 IEEE Transportation Electrification Conference and Expo (ITEC), Dearborn, MI, USA, 18–20 June 2012; pp. 1–5. [CrossRef]
65. Di Rienzo, R.; Baronti, F.; Vellucci, F.; Cignini, F.; Ortenzi, F.; Pede, G.; Saletti, R. Experimental analysis of an electric minibus with small battery and fast charge policy. In Proceedings of the 2016 International Conference on Electrical Systems for Aircraft, Railway, Ship Propulsion and Road Vehicles and International Transportation Electrification Conference, ESARS-ITEC, Toulouse, France, 2–4 November 2016. [CrossRef]
66. Rivera-Barrera, J.P.; Muñoz-Galeano, N.; Sarmiento-Maldonado, H.O. SoC Estimation for Lithium-ion Batteries: Review and Future Challenges. *Electronics* **2017**, *6*, 102. [CrossRef]
67. Xiong, R.; Cao, J.; Yu, Q.; He, H.; Sun, F. Critical Review on the Battery State of Charge Estimation Methods for Electric Vehicles. *IEEE Access* **2018**, *6*, 1832–1843. [CrossRef]
68. Berecibar, M.; Gandiaga, I.; Villarreal, I.; Omar, N.; Van Mierlo, J.; Van den Bossche, P. Critical review of state of health estimation methods of Li-ion batteries for real applications. *Renew. Sustain. Energy Rev.* **2016**, *56*, 572–587. [CrossRef]
69. Ungurean, L.; Cârstoiu, G.; Micea, M.V.; Groza, V. Battery state of health estimation: A structured review of models, methods and commercial devices. *Int. J. Energy Res.* **2017**, *41*, 151–181. [CrossRef]

70. Lin, Q.; Wang, J.; Xiong, R.; Shen, W.; He, H. Towards a smarter battery management system: A critical review on optimal charging methods of lithium ion batteries. *Energy* **2019**, *183*, 220–234. [CrossRef]
71. Hillers, A.; Christen, D.; Biela, J. Design of a Highly Efficient Bidirectional Isolated LLC Resonant Converter. In Proceedings of the 2012 15th International Power Electronics and Motion Control Conference (EPE/PEMC), Novi Sad, Serbia, 4–6 September 2012.
72. Gadelrab, R.; Yang, Y.; Li, B.; Lee, F.C.; Li, Q. High-Frequency High-Density Bidirectional EV Charger. In Proceedings of the IEEE Transportation Electrification Conference and Expo (ITEC) 2018, Long Beach, CA, USA, 13–15 June 2018.
73. Li, B.; Li, Q.; Lee, F.C. A Novel PCB Winding Transformer with Controllable Leakage Integration for a 6.6 kW 500 kHz High Efficiency High Density Bi-Directional On-Board Charger. In Proceedings of the 2017 IEEE Applied Power Electronics Conference and Exposition (APEC), Tampa, FL, USA, 26–30 March 2017; pp. 2917–2924.
74. Li, B.; Li, Q.; Lee, F.C.; Liu, Z.; Yang, Y. A High-Efficiency High-Density Wide-Bandgap Device-Based Bidirectional On-Board Charger. *IEEE J. Emerg. Sel. Top. Power Electron.* **2018**, *6*, 1627–1636. [CrossRef]
75. Ejuri, J. How to Compare the Figure of Merit (FOM) of MOSFETs. In *Infineon Application Note*; 2003; Available online: http://cdn14.21dianyuan.com (accessed on 2 October 2019).
76. Saponara, S.; Fanucci, L.; Bernardo, F.; Falciani, A. Predictive diagnosis of high-power transformer faults by networking vibration measuring nodes with integrated signal processing. *IEEE Trans. Instrum. Meas.* **2016**, *65*, 1749–1760. [CrossRef]
77. Saponara, S.; Bacchillone, T. Network architecture, security issues, and hardware implementation of a home area network for smart grid. *J. Comput. Netw. Commun.* **2012**, *2012*, 534512. [CrossRef]

applied sciences

MDPI

Article

Research on Virtual Inductive Control Strategy for Direct Current Microgrid with Constant Power Loads

Zhiping Cheng, Meng Gong, Jinfeng Gao, Zhongwen Li * and Jikai Si

School of Electrical Engineering, Zhengzhou University, Zhengzhou 450001, China; zpcheng@zzu.edu.cn (Z.C.); gongmeng8036@163.com (M.G.); jfgao@zzu.edu.cn (J.G.); sijikai527@126.com (J.S.)

* Correspondence: lzw@zzu.edu.cn

Received: 16 September 2019; Accepted: 18 October 2019; Published: 20 October 2019

check for updates

Abstract: In order to improve the stability of direct current (DC) microgrid with constant power loads, a novel virtual inductive approach is proposed in this paper. It is known that the negative impedance characteristic of constant power loads will lead to DC bus voltage fluctuation, which will be more serious when they integrate into the DC microgrid though a large transmission line inductive. For the convenience of analysis, a simplified circuit model of the system is obtained by modeling the distributed resources. Unlike the existing control strategies, the proposed control strategy constructs a negative inductance link, which helps to counteract the negative effects of the line inductive between the power source and the transmission line. Detailed performance comparison of the proposed control and virtual capacitance are implemented through MATLAB/simulink simulation. Moreover, the improved performance of the proposed control method has been further validated with several detailed studies. The results demonstrate the feasibility and superiority of the proposed strategy.

Keywords: direct current (DC) microgrid; stability; constant power loads; large transmission line inductive; simplified circuit model; virtual negative inductance

---

1. Introduction

There is no fluctuation of reactive power in the microgrid, and the stability of the direct current (DC) microgrid only depends on the stable operation of the bus voltage [1,2]. When the DC microgrid is operating in island mode, without the supporting of a large power grid, the instability of bus voltage will be more serious [3]. From the structural point of view, the island DC microgrid is a typical multi-source heterogeneous system containing multiple types of micro-resources and loads [4,5]. The stability is not only affected by the distributed resources, but also subjected to load, especially constant power loads (CPLs). Thus, how to eliminate the influence of the CPLs characteristics caused by the load converter and primary load is a technical challenge to ensure the stability of a DC microgrid [6].

To overcome the problem of voltage fluctuation caused by CPLs, the instability mechanism of CPLs has been widely studied and several solutions have been proposed [7,8]. In the past few years, some researchers have tried to solve such instability phenomena by a passive method [9–11]. Generally speaking, this method requires additional physical dampers. In [9], an LC (Inductive Capacitance) filter is added to the system. Although it shows a better effect, the system is based on an ideal model, which is different from the actual situation. And the response speed of the system is also affected to some extent. Based on the same idea, a physical filter containing multiple capacitors and resistors is applied to eliminate the impact of constant power loads [11]. However, the use of multiple resistors will make the power loss greater. Previous studies indicate that the power loss caused by passive damping makes it not suitable for DC microgrids. Therefore, the active control strategy gets more attention. This kind of method will not bring any additional physical dampers to the system, hence alleviating the system weight and efficiency problem. This is more in line with the requirements of efficient energy utilization in a DC microgrid. Furthermore, because of the advantages of distributed control, droop control has been widely used as an important form of active control [12–15].

Active control method can achieve the same effect as passive control strategy, but it will not cause additional power loss. Guo et al. added a low-pass filter in series to droop control, which can increase the output impedance of the converter in the low frequency range [16]. Thus, the bus voltage fluctuation caused by CPLs can be effectively alleviated. However, this research is limited to the introduction of filter designs. Furthermore, for solving the negative effects of CPLs, Wu et al. described a virtual phase-lead impedance stability control strategy [17]. The strategy was used to adjust the output impedance of CPLs to be positive. But it was not applied to an island mode. And from the point of inertia, a corresponding virtual capacitance control strategy was established by Zhu et al. [18]. And due to the increase of inertia by virtual capacitance, the robustness of system is increased in the face of CPLs. However, because the "plug-and-play" characteristics of distributed resources, new control strategies need to be designed for the access of new micro resources, which obviously lacks flexibility. In addition, other methods are also committed to using active control methods to improve system stability [19,20]. Among various approaches investigated in the literature, there is a deficiency that the influence of transmission line is not included in these study. This is imperfect, because the transmission line is also a major factor affecting the stability of the system, especially the inductance component of the transmission line [21,22].

To address the above problem, a virtual negative inductance control strategy is proposed in this paper. On the basis of droop control, a virtual inductive is constructed, which operates as a negative inductive link and counteracts the inductive in transmission line. Thus, the negative influence of line inductance is effectively solved. This paper makes the following contributions:

(1) Through the modeling of micro-sources and loads, the system can realize the conversion from a multi-dimensional to a low-dimensional system.
(2) An innovative solution of system instability is proposed from the perspective of line inductance. Compared with the virtual capacitance strategy in [18], the method proposed in this paper will be more smooth and fast in the process of adjustment.
(3) The other salient feature of this paper is that starting from the transmission line, the strategy can be applied to a wide range of situations.

The remainder of the paper is organized as follows: The configuration and modeling of the DC microgrid is introduced in Section 2, and the simplified circuit model of the system is shown. Section 3 presents the design of the virtual inductive control strategy. In Section 4, simulation results are used to verify the superiority of the strategy by comparison and different case studies. Finally, Section 5 summarizes this paper and draws conclusions.

2. Structure and Modeling of Island DC Microgrid

The structure of an island DC microgrid is depicted in Figure 1, which mainly composes of four parts: (1) distributed resources (PV and wind turbine); (2) eEnergy storage unit; (3) various types of load (mainly divided into CPLs and resistive load); and (4) line impedance. Subsystems in the DC microgrid can also be divided into power control unit and constant voltage control unit according to their control mode [19]. The purpose of power control is to ensure that distributed generators inject constant power or maximum power into the system. In this paper, MPPT (maximum power point tracking) control was used for distributed generators such as wind turbines and PVs to achieve maximum power utilization. The constant voltage unit is aimed at stabilizing bus voltage [23]. For the energy storage unit, a droop control method is applied to the DC bus voltage [24].

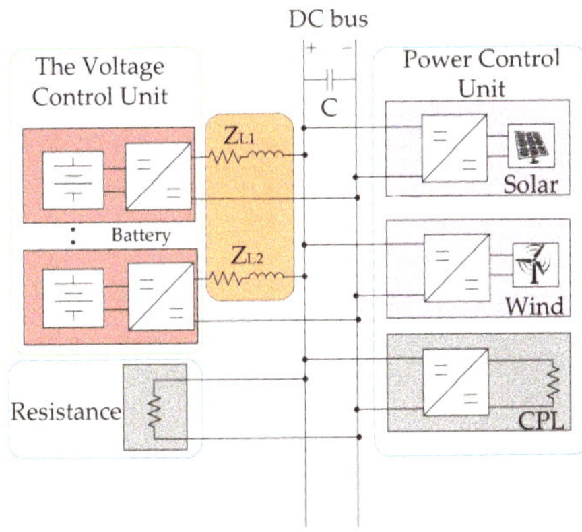

Figure 1. Typical configuration of an island direct current (DC) micro grid.

2.1. Modeling of Distributed Power Unit

The resources are connected to the DC microgrid through power converters, whose control dictates their behavior. Resources under MPPT try to maximize the injection of available power to the DC microgrid regardless of the network status. Under constant weather conditions, the power converter of distributed generation can be modeled as constant power sources (CPSs) when viewed by the bus terminals [5]. This means that despite of the variation on the bus voltage, the current provided by the power converter adapts to keep injecting constant power. Such distributed resources can be considered to be a type of CPLs with a negative power consumption, and both constant power loads and constant power sources can be totally modeled as an ideal constant power model, as illustrated in Figure 2a (taking the photovoltaic as an example). As CPLs and CPSs behave in the same way, they can be modeled as lumped CPLs that demand an equivalent constant power P given by (1). And based on [21], energy storage units based on the droop control are modeled as an ideal voltage source with series resistance, as illustrated in Figure 2b.

$$P = P_{CPLs} + P_{CPSs} P_{CPLs} > 0 \text{ and } P_{CPSs} < 0, \tag{1}$$

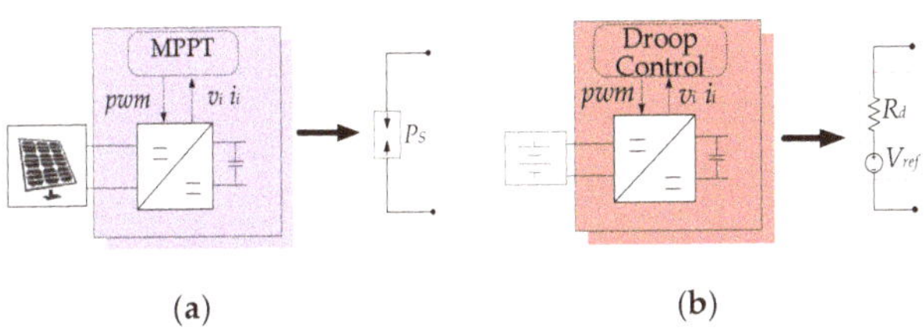

Figure 2. Distributed Resources model. (**a**) Maximum Power Point Tracking (MPPT) control. (**b**) Droop control.

2.2. Dimension Reduction Modeling of Multi-Voltage Control Unit under Droop Control

On the basis of the work done above, the multi-branch combination model of DC microgrids with multiple voltage control units was obtained, as illustrated in Figure 3.

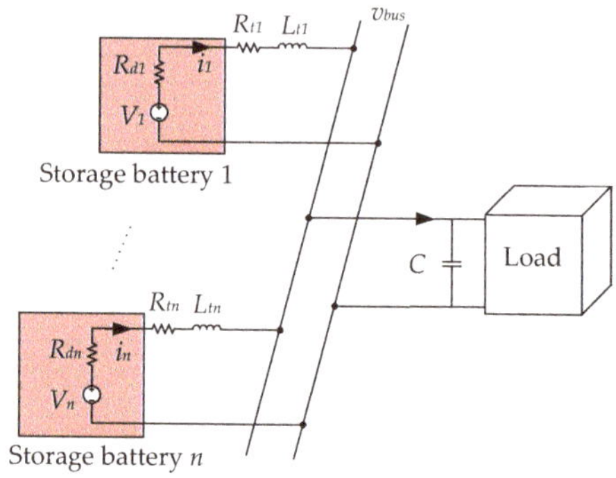

Figure 3. Multi-branch combination model of n droop control resources.

V_i is the source voltage and v_{bus} is the voltage of the dc bus. R_{di} is the droop coefficient of the ith droop controller, R_{ti} and L_{ti} are the line resistance and inductance from ith energy storage unit to the bus, respectively. C is the equivalent capacitance on the bus [25]. The droop resistance and line resistance in series of the ith control unit can be represented by:

$$R_i = R_{di} + R_{ti}, \tag{2}$$

Assume that the value of droop resistance is much greater than the line resistance, Equation (2) can be rewritten as:

$$R_i \approx R_{di}, \tag{3}$$

As a complex multi-dimensional system, the DC microgrid power system is not conducive to analysis [26]. It is necessary to simplify the n-dimensional equation of representative DC microgrid system. Each distributed unit can be modeled as a differential equation in (4).

$$\begin{aligned} \frac{di_1}{dt} &= \frac{1}{L_{t1}}(V_1 - v_{bus}) - \frac{R_{d1}}{L_{t1}} i_1, \\ \frac{di_2}{dt} &= \frac{1}{L_{t2}}(V_2 - v_{bus}) - \frac{R_{d2}}{L_{t2}} i_2, \\ &\vdots \\ \frac{di_n}{dt} &= \frac{1}{L_{tn}}(V_n - v_{bus}) - \frac{R_{dn}}{L_{tn}} i_n. \end{aligned} \tag{4}$$

Furthermore, the total current i_s provided by the resources can be denoted by

$$i_s = i_1 + i_2 + \cdots + i_n, \tag{5}$$

Therefore, the sum of the n differential equations in (4) can be simplified to

$$\frac{di_s}{dt} = \left(\sum_{i=1}^{n} \frac{1}{L_{ti}} \right) (V_i - v_{bus}) - \frac{R_{di}}{L_{ti}} i_s, \tag{6}$$

In reality, multiple identical energy storage units often exist in the form of clusters [27], and the overall power supply capacity is evenly divided into individual battery unit. Assuming that the specifications of each energy storage unit and the line length to the bus are the same. Thus, the following equation is obtained

$$\begin{aligned} V_{ref} &= V_1 = V_2 = \cdots = V_n, \\ R_{d1} &= R_{d2} = \cdots = R_{dn}, \\ L_{t1} &= L_{t2} = \cdots = L_{tn}. \end{aligned} \tag{7}$$

Multiplying both sides of (6) by

$$L_t = \frac{1}{\sum_{i=1}^{n} \frac{1}{L_{tn}}}, \tag{8}$$

The system is simplified from multiple differential equations to a simple differential equation in (9). Thus, the conversion from multi-dimensional to a low-dimensional is realized

$$L_t \frac{di_s}{dt} = \left(V_{ref} - v_{bus}\right) - L_t \frac{R_{dn}}{L_{tn}} i_s, \tag{9}$$

And under the condition:

$$R_d = L_t \frac{R_{dn}}{L_{tn}}, \tag{10}$$

From (9), the equivalent steady-state model of the DC microgrid system can be obtained, as illustrated in Figure 4.

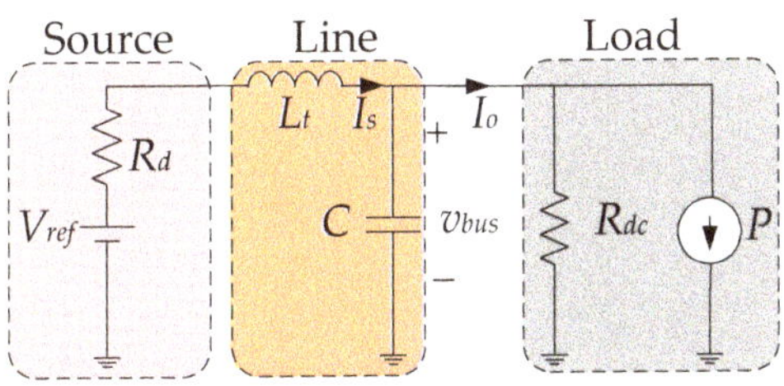

Figure 4. Equivalent steady-state model of a DC microgrid.

where V_{ref} is the reference voltage of the bus. Resistance R_d is the droop control coefficient. P, R_{dc} correspond to the equivalent power of CPLs and the resistive load, respectively. The corresponding whole system parameters are listed in Table 1.

Table 1. Parameters of DC microgrid system.

Parameter	Magnitude
Input Voltage V_S	100 v
Bus Voltage V_{bus}	200 v
Filter Inductor L/R	13.68 mH/0.4Ω
Bus Capacitance C	840 μF
Switching frequency f_s	10 kHz
Droop coefficient R_d	0.5
Resistive load R_{dc}	100 Ω
Line inductor L_t	0.5 mH
Proportional Coefficient of Current Controller k_{pi}	0.5
Integral Coefficient of Current Controller k_{ii}	100
Proportional Coefficient of Voltage Controller k_{pv}	1.2
Integral Coefficient of Voltage Controller k_{iv}	250
Filter Coefficient T	0.0005
Droop Coefficient R_d	0.5
virtual inductive L_d	5

3. Virtual Inductive Control Strategy

3.1. *Definition of Instability*

For CPLs, the input power is equal to the required power of the CPLs. And the power characteristics can be given by

$$P = V_{bus} i_{load}, \tag{11}$$

And (12) can be obtained by small signal analysis.

$$\Delta i_{load} = -\frac{P}{V_{bus}^2}\Delta v_{bus}, \tag{12}$$

where Δi_{load} and Δv_{bus} are the perturbation of load current and bus voltage, respectively. V_{bus} is the steady-state bus voltage value. It can be noted from (12) that the input current of CPLs varies inversely with bus voltage at a multiple of P/V_{bus}^2. Thus, the rise (decrease) of voltage corresponds to the decrease (rise) of current. That is why the CPLs cause instability, also known as negative impedance characteristics. It is noteworthy in (12) that the negative impedance characteristic effect of the CPLs on the bus voltage can be eliminated by offsetting the current change. The virtual inductive control strategy of this paper will be introduced in detail below.

3.2. Virtual Inductance Control Strategy

Voltage and current double closed-loop control is a common control method [12,18]. And transfer function of voltage and current double closed-loop control of DC microgrid given by

$$G_{CLi}(s) = \frac{\left(k_{pi}s + k_{ii}\right)V_0}{Ls^2 + k_{pi}V_0 s + k_{ii}V_0}, \tag{13}$$

$$G_{CLv}(s) = \frac{\left(k_{pv}s + k_{iv}\right)(1-D)}{Cs^2 + k_{pv}(1-D)s + k_{iv}(s)}, \tag{14}$$

k_{pi}, k_{ii} are the proportional and integral parameters of current loop, respectively. And k_{pv}, k_{iv} are the proportional and integral parameters of voltage loop, respectively. According to the selection rules of controller parameters in [16], the specific PI controller parameters of this paper given in Table 1. The dynamic characteristics of the current inner loop can be equivalent to proportional gain one when calculating the voltage outer loop transfer function. The small signal block diagram of the droop control strategy can be derived, as illustrated in Figure 5 [28].

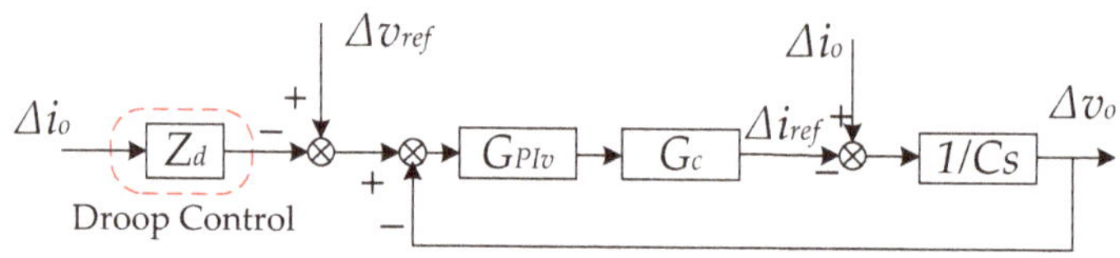

Figure 5. Small signal control block diagram of droop control.

where Z_d is the droop control, $\Delta v_o, \Delta v_{ref}, \Delta i_o, \Delta i_{ref}$ are the perturbation of the output voltage, the reference voltage, output current, and the output reference current of double closed-loop controller, respectively. $G_{PIv}(s)$ is the voltage control loop transfer function and $G_c(s)$ is the current inner loop controller closed-loop transfer function. The output voltage perturbation
$varDeltav_o$ can be given by:

$$varDeltav_o = \frac{G_{PIv(s)}G_{c(s)}}{Cs + G_{PIv}(s)G_C(s)} varDeltav_{ref} - \frac{1 + Z_d G_{PIv}(s)G_c(s)}{Cs + G_{PIv}(s)G_c(s)} varDeltai_o, \tag{15}$$

In (15), the change of $varDeltav_{ref}$ and $varDeltai_0$ have an effect on the bus voltage. It is generally considered that the reference voltage of the system is unique and the fluctuation of the output voltage is determined by the change of the output current. Therefore, reasonable design of control strategy can reduce the influence of the latter. The definition is as follows

$$1 + Z_d G_{PIv}(s) G_c(s) = 0, \tag{16}$$

And Z_d can be rewritten as

$$Z_d = -L_d s \frac{1}{Ts+1}, \tag{17}$$

where $L_d = 1/k_{iv}$, and $T = k_{pv}/k_{iv}$. Formula (18) can be considered as a form of multiplication of a coefficient $-L_d$ differential link with a low-pass filter. According to the actual situation, the value of $-L_d$ is chosen as shown in Table 1. On the basis of the traditional droop control, the control form is expressed as follows

$$Z_d = R_d - L_d s, \tag{18}$$

And an improved droop controller block diagram is depicted in Figure 6.

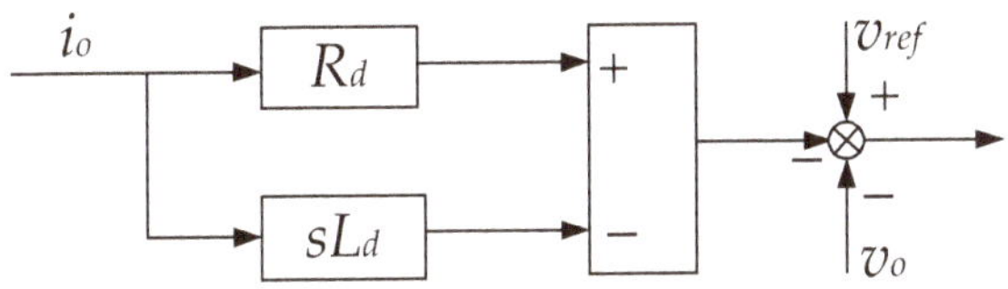

Figure 6. Improved droop controller.

As shown in Figure 7, with the proposed control strategy, an equivalent virtual control link is added between the source and the transmission line, which is equivalent to a negative inductance link in the system, helps to offset the influence of the line inductance in the system.

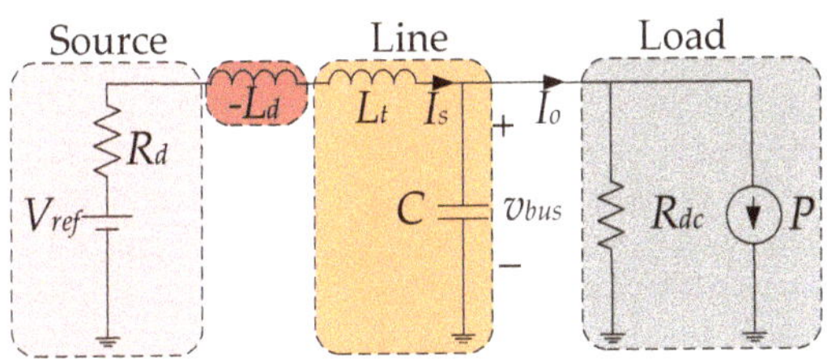

Figure 7. Virtual negative inductive reactance control strategy.

The control parameters of system are shown in Table 1. Where the value of droop coefficient depends on the maximum voltage deviation
$varDeltav_{0max}$ and the rated output current i_0. The maximum allowable voltage deviation is ±5% of the rated bus voltage. R_d can be derived as:

$$R_d = \frac{varDeltav_{0max}}{i_0}, \tag{19}$$

4. Simulation Verification

In this section, simulation results in Matlab/Simulink are shown to validate the proposed control strategy. The changes in bus voltage, inductor current, and output current of the DC microgrid before and after the application of the control strategy are shown in Figure 8. Initially, the traditional droop control was applied. Due to the influence of CPLs, the system suffers oscillation. At t = 0.5s, the control strategy is switched from the traditional droop control to the proposed control strategy. Obviously, the DC microgrid bus voltage can quickly be restored to a stable state. Compared with the virtual capacitance control strategy based on droop control (the gray curve in the figure) in [18], the proposed control strategy in this paper can reduce the adjustment time and make the system reach the balance faster. In addition, the control strategy can reduce the bus voltage fluctuation during the regulation process. Overall, this strategy will have more advantages in improving stability. To further verify the effectiveness of the proposed strategy, the following tests are carried out with the change of droop coefficients, line lengths, and operating points.

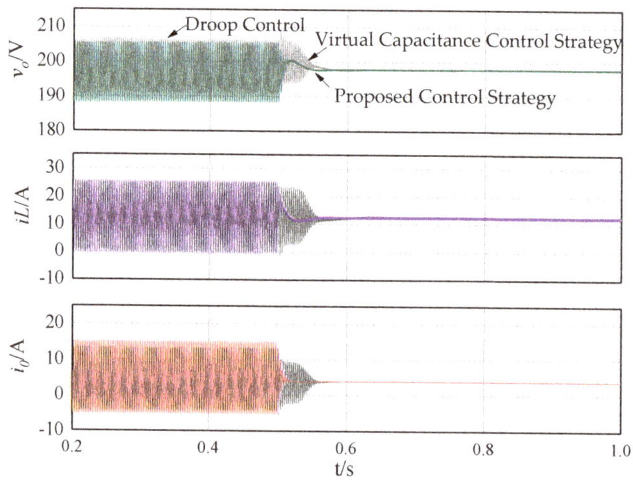

Figure 8. Simulation results of a DC microgrid system with the proposed control strategy.

4.1. Change of Droop Coefficients

DC microgrids can coordinate the output of resources by adjusting the droop coefficient, which means that the system may operate under different droop coefficients due to different requirements [29]. Therefore, the stable operation of the system under different droop coefficients must be ensured. Figure 9 shows the simulation result of the proposed control strategy under different droop coefficients. It can be seen that the system remained stable whether the droop coefficient was switched to 0.3 or 0.7. Thus, it is evident that the control strategy can ensure the system stability under different droop coefficients.

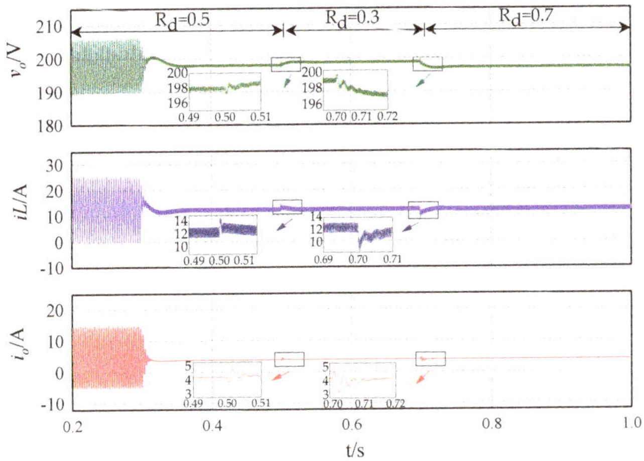

Figure 9. Simulation results of DC microgrid system with the change droop coefficient.

4.2. Change of Line Lengths

The system needs to meet the energy requirements of the load, but the distance between different loads and the bus are different. It is generally considered that the value of the droop coefficient is much larger than the line resistance and only the inductive component in the line is considered in this paper. The inductance of the line is proportional to the length. Hence, the variable considered in this section is the line length. As shown in Figure 10, the line length varies following the sequence of 1.070 km, 0.535 km and 0.749 km. It can be seen that no matter the length of the line increases or decreases, the DC microgrid can keep stable after applying the proposed control strategy. Thus, the effectiveness of the proposed stabilization strategy is proved when the system line length varies.

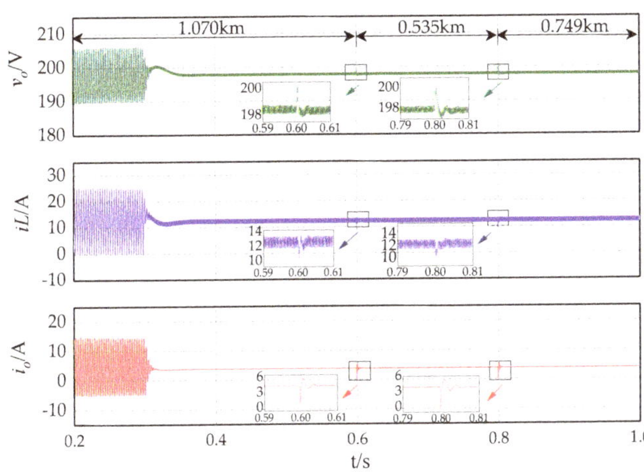

Figure 10. Simulation results of DC microgrid system with the change of line length.

4.3. Change of Operating Points

The load in the DC microgrid is an important part of the system and one of the main factors affecting the stability of the system. Instability problems are more likely to occur when load changes [30,31]. Thus, the effectiveness of the proposed control strategy was verified by changing the power of CPLs.

The load doubled at 0.5 s, tripled at 0.7 s, and recovered to the initial at 0.9 s, as illustrated in Figure 11. It can be seen that the system operates steadily whether the power increases or decreases. It proves that the control strategy can improve the system robustness in case of load variation.

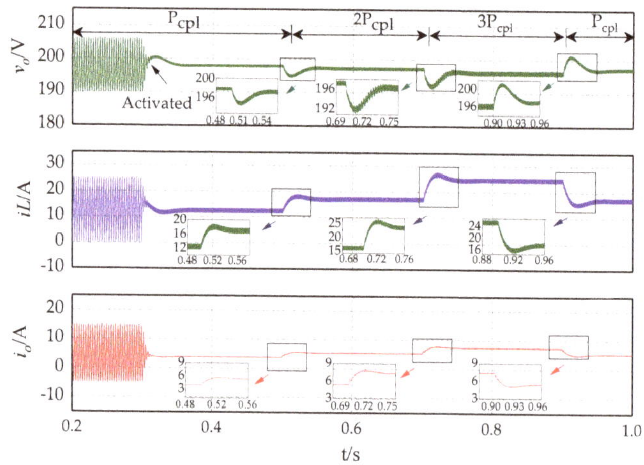

Figure 11. Simulation results of a DC microgrid system with the change of operation points.

5. Conclusions

A virtual negative inductance control strategy for DC microgrids with constant power loads is proposed in this paper. The particular is that this strategy improves the stability of the system from the point of line inductance. And the main contributions of this paper lie in the following:

(1) Complex multi-dimensional DC microgrid systems are not conducive to analysis. So, a concise system model was obtained by simplifying the distributed generation model and the multi-branch model. Moreover, the model takes into account multiple types of power supply and loads in a DC microgrid system, which is closer to the actual situation.

(2) The instability caused by constant power loads can be attributed to the change of output current Δi_0. The expression of the control strategy in this paper was deduced to solve the problem of current effect. And the form of control strategy was equivalent to adding a negative inductance link to the system model.

(3) The effectiveness of this strategy was verified and compared with the existing virtual capacitance control strategy. The results show that this strategy can effectively solve the instability of DC microgrid. It also proves that the proposed strategy has faster and smoother regulation process when compared with the virtual capacitance strategy.

In this paper, the stability of a single DC microgrid is studied. For future work, the stability of a DC microgrid group will be of concern. The power generation variation of renewable energy based distributed generators will also be considered. And more detailed research will be carried out by considering more detailed transient models of distributed generation and load. Besides, we also intend to extend the proposed control strategy to a hybrid alternating current (AC)/DC microgrid.

Author Contributions: Conceptualization, Z.C. and M.G.; methodology, G.M. and Z.L.; software, Z.C., M.G. and Z.L.; validation, Z.C. and M.G. and J.G.; formal analysis, Z.C. and M.G.; investigation, M.G.; resources, Z.L.; data curation, J.G.; writing—original draft preparation, M.G.; writing—review and editing, Z.C. and M.G.; visualization, Z.C. and M.G.; supervision, J.G., Z.L. and J.S.; project administration, Z.L.; and funding acquisition, M.G.

Funding: This work was supported by the National Natural Science Foundation of China (61803343), the China Postdoctoral Science Foundation (2018M630835), and the Key Projects of Higher Education of Henan Province (18A470015,18A470017, and 19A120012).

Conflicts of Interest: The authors declare no conflict of interest.

Abbreviations

The following abbreviations are used in the manuscript:

DC	Direct Current
AC	Alternating Current
CPLs	Constant Power Loads
CPSs	Constant Power Sources
MPPT	Maximum Power Point Tracking
PV	Photovoltaic

References

1. Bayati, N.; Hajizadeh, A.; Soltani, M. Protection in DC microgrids: A comparative review. *IET Smart Grid* **2018**, *1*, 66–67. [CrossRef]
2. Naderi, Y.; Hosseini, S.H.; Ghassem Zadeh, S.; Mohammadi-Ivatloo, B.; Vasquez, J.C.; Guerrero, J.M. An overview of power quality enhancement techniques applied to distributed generation in electrical distribution networks. *Renew. Sustain. Energy Rev.* **2018**, *93*, 201–214. [CrossRef]
3. Kamel, R.M.; Chaouachi, A.; Nagasaka, K. Detailed Analysis of Micro-Grid Stability during Islanding Mode under Different Load Conditions. *Engineering* **2011**, *3*, 508–516. [CrossRef]
4. Shivam; Dahiya, R. Stability analysis of islanded DC microgrid for the proposed distributed control strategy with constant power loads. *Comput. Electr. Eng.* **2018**, *70*, 151–162. [CrossRef]
5. Tahim, A.P.N.; Pagano, D.J.; Lenz, E.; Stramosk, V. Modeling and Stability Analysis of Islanded DC Microgrids Under Droop Control. *IEEE Trans. Power Electron.* **2015**, *30*, 4597–4607. [CrossRef]
6. Su, M.; Liu, Z.; Sun, Y.; Han, H.; Hou, X. Stability Analysis and Stabilization Methods of DC Microgrid with Multiple Parallel-Connected DC–DC Converters Loaded by CPLs. *IEEE Trans. Smart Grid* **2018**, *9*, 132–142. [CrossRef]
7. Pakdeeto, J.; Areerak, K.; Areerak, K. Large-signal model of DC micro-grid systems feeding a constant power load. In Proceedings of the 2017 International Electrical Engineering Congress (iEECON), Pattaya, Thailand, 8–10 March 2017; pp. 1–4.
8. Liu, Z.; Su, M.; Sun, Y.; Han, H.; Hou, X.; Guerrero, J.M. Stability analysis of DC microgrids with constant power load under distributed control methods. *Automatica* **2018**, *90*, 62–72. [CrossRef]
9. Cespedes, M.; Xing, L.; Sun, J. Constant-Power Load System Stabilization by Passive Damping. *IEEE Trans. Power Electron.* **2011**, *26*, 1832–1836. [CrossRef]
10. Céspedes, M.; Beechner, T.; Xing, L.; Sun, J. Stabilization of constant-power loads by passive impedance damping. In Proceedings of the 2010 Twenty-Fifth Annual IEEE Applied Power Electronics Conference and Exposition (APEC), Palm Springs, CA, USA, 21–25 February 2010; pp. 2174–2180.
11. Mayo-Maldonado, J.C.; Rapisarda, P. A systematic approach to constant power load stabilization by passive damping. In Proceedings of the IEEE Conference on Decision and Control, Osaka, Japan, 15–18 December 2015; pp. 1346–1351.
12. Kim, H.-J.; Lee, Y.-S.; Kim, J.-H.; Han, B.-M. Coordinated droop control for stand-alone DC micro-grid. *J. Electr. Eng. Technol.* **2014**, *9*, 1072–1079. [CrossRef]
13. Ko, B.-S.; Lee, G.-Y.; Choi, K.-Y.; Kim, R.-Y. A Coordinated Droop Control Method Using a Virtual Voltage Axis for Power Management and Voltage Restoration of DC Microgrids. *IEEE Trans. Ind. Electron.* **2019**, *66*, 9076–9085. [CrossRef]
14. Cheng, Z.; Li, Z.; Liang, J.; Gao, J.; Si, J.; Li, S. Distributed Economic Power Dispatch and Bus Voltage Control for Droop-Controlled DC Microgrids. *Energies* **2019**, *12*, 1400. [CrossRef]
15. Lu, X.; Guerrero, J.M.; Sun, K.; Vasquez, J.C. An Improved Droop Control Method for DC Microgrids Based on Low Bandwidth Communication with DC Bus Voltage Restoration and Enhanced Current Sharing Accuracy. *IEEE Trans. Power Electron.* **2014**, *29*, 1800–1812. [CrossRef]

16. Guo, L.; Zhang, S.; Li, X.; Li, Y.W.; Wang, C.; Feng, Y. Stability Analysis and Damping Enhancement Based on Frequency-Dependent Virtual Impedance for DC Microgrids. *IEEE J. Emerg. Sel. Top. Power Electron.* **2017**, *5*, 338–350. [CrossRef]
17. Wu, W.; Chen, Y.; Zhou, L.; Zhou, X.; Yang, L.; Dong, Y.; Xie, Z.; Luo, A. A Virtual Phase-Lead Impedance Stability Control Strategy for the Maritime VSC–HVDC System. *IEEE Trans. Ind. Inform.* **2018**, *14*, 5475–5486. [CrossRef]
18. Zhu, X.; Xie, Z.; Jing, S.; Ren, H. Distributed virtual inertia control and stability analysis of dc microgrid. *IET Gener. Transm. Distrib.* **2018**, *12*, 3477–3486. [CrossRef]
19. Hosseinipour, A.; Hojabri, H. Virtual inertia control of PV systems for dynamic performance and damping enhancement of DC microgrids with constant power loads. *IET Renew. Power Gener.* **2018**, *12*, 430–438. [CrossRef]
20. Huangfu, Y.; Pang, S.; Nahid-Mobarakeh, B.; Guo, L.; Rathore, A.K.; Gao, F. Stability Analysis and Active Stabilization of On-board DC Power Converter System with Input Filter. *IEEE Trans. Ind. Electron.* **2018**, *65*, 790–799. [CrossRef]
21. Anand, S.; Fernandes, B.G. Reduced-Order Model and Stability Analysis of Low-Voltage DC Microgrid. *IEEE Trans. Ind. Electron.* **2013**, *60*, 5040–5049. [CrossRef]
22. Tah, A.; Das, D. Analysis on different aspects of interconnected DC microgrids. In Proceedings of the 2016 IEEE International Conference on Power Electronics, Drives and Energy Systems (PEDES), Trivandrum, India, 14–17 December 2016; pp. 1–6.
23. Gunasekaran, M.; Ismail, H.M.; Chokkalingam, B.; Mihet-Popa, L.; Padmanaban, S. Energy Management Strategy for Rural Communities' DC Micro Grid Power System Structure with Maximum Penetration of Renewable Energy Sources. *Appl. Sci.* **2018**, *8*, 585. [CrossRef]
24. Marcon, P.; Szabo, Z.; Vesely, I.; Zezulka, F.; Sajdl, O.; Roubal, Z.; Dohnal, P. A Real Model of a Micro-Grid to Improve Network Stability. *Appl. Sci.* **2017**, *7*, 757. [CrossRef]
25. Xu, L.; Chen, D. Control and Operation of a DC Microgrid with Variable Generation and Energy Storage. *IEEE Trans. Power Deliv.* **2011**, *26*, 2513–2522. [CrossRef]
26. Wei, Y.; Luo, Q.; Chen, S.; Huang, J.; Zhou, L. DC current bus distributed power system and its stability analysis. *IET Power Electron.* **2019**, *12*, 458–464. [CrossRef]
27. Chang, Y.C.; Chang, H.C.; Huang, C.Y. Design and Implementation of the Battery Energy Storage System in DC Micro-Grid Systems. *Energies* **2018**, *11*, 1566. [CrossRef]
28. Li, Y.; Fan, L. Stability Analysis of Two Parallel Converters with Voltage–Current Droop Control. *IEEE Trans. Power Deliv.* **2017**, *32*, 2389–2397. [CrossRef]
29. Mokhtar, M.; Marei, M.I.; El-Sattar, A.A. An Adaptive Droop Control Scheme for DC Microgrids Integrating Sliding Mode Voltage and Current Controlled Boost Converters. *IEEE Trans. Smart Grid* **2019**, *10*, 1685–1693. [CrossRef]
30. Moussa, S.; Ghorbal, M.J.; Slama-Belkhodja, I.; Martin, J.; Pierfederici, S. DC bus voltage instability detection and stabilization under constant power load variation. In Proceedings of the 2018 9th International Renewable Energy Congress (IREC), Hammamet, Tunisia, 20–22 March 2018; pp. 1–6.
31. Xiao, B.; Dai, D.; Zhang, B.; Han, Y. Bifurcation and its analysis of DC Micro-Grid under abrupt load change. *J. Electr. Eng.* **2016**, *31*, 131–141.

applied sciences

MDPI

Article

The Improvement of an Islanding Detection Method Based on the Perturbation Signal in Case of a Multi-Photovoltaic Operation

Thanh Son Tran [1,2,*], **Duc Tuyen Nguyen** [2] and **Goro FUJITA** [3]

1 Graduate School of Engineering and Science, Shibaura Institute of Technology, Tokyo 135-8548, Japan
2 Department of Power System, Hanoi University of Science and Technology, Hanoi 11615, Vietnam; tuyen.nguyenduc@hust.edu.vn
3 Department of Electrical Engineering, Shibaura Institute of Technology, Tokyo 135-8548, Japan; gfujita@sic.shibaura-it.ac.jp
* Correspondence: na16502@shibaura-it.ac.jp; Tel.: +81-806-188-9630

Received: 22 August 2019; Accepted: 24 September 2019; Published: 27 September 2019

check for updates

Abstract: Islanding phenomenon is one of the consequences of the emergence and development of microgrids in the power system. Injected signal cancellation is a common problem in a multi-distributed generation that has a significant influence on active islanding detection methods. In this study, this issue was analyzed by injecting a perturbation signal in the multi-photovoltaic system. Furthermore, the promising solution to eliminate injected signal cancellation was proposed in this paper. The solution was validated through mathematical explanations and simulation results.

Keywords: islanding detection; DC grid; active islanding detection method; injected signal cancellation; multi-distributed generations; perturbation signal; photovoltaic system

1. Introduction

In recent years, the penetration of dispersed generations (DGs) such as photovoltaic and wind energy in electricity systems has significantly increased. Some of the distributed generations are natively DC. The developments of DC loads and storage technologies have also considerably increased [1]. The DC grid has much more advantages such as higher system efficiency, easier integration and lower losses due to reduced conversion devices than the AC grid [2–5]. Therefore, the DC grid has been rapidly increasing in the power system. One of the major challenges with DC grid-connected DG is the detection of the unintentional islanding phenomenon. According to the IEEE 929-2000 standard [6], islanding problem occurs when the distributed network isolated from the main utility and the loads are supplied by DGs only. Based on this standard, the islanding phenomenon must be detected within 2 s to prevent damage to electrical devices and ensure the safety of maintenances. Nowadays, islanding detection has received considerable attention through recent studies.

In [7], a novel islanding detection method based on two additional compensators and positive feedback of active and reactive power control loops. The method does not have a non-detection zone, no affect on power quality and normal operation of synchronous DG. Because of it need additional compensators, it may cause the method to complicate and costly.

A novel islanding detection method for micro-grid was proposed in [8]. The method can detect islanding phenomenon in hardest condition (low active and reactive power mismatches) by using a differential morphological filter (a dilation–erosion differential filter (DED)) of the RMS signal (DEDFOR) at the point of common coupling. The method has a small non-detection zone (NDZ), fast detection time, and high accuracy. However, the method is used for double-fed induction generators (DFIGs) system.

A new passive islanding detection includes in Smart Islanding Detector (SmartID) was proposed in [9]. It combines with classical voltage and frequency relays to make a better anti-islanding protection system. The method is used to backup in case of the classical method fail. However, the combination of SmartID and classical relays also complicated.

In [10], a new islanding detection strategy for low-voltage inverter-interfaced microgrids was presented. This strategy was based on adaptive neuro-fuzzy inference system (ANFIS). The ANFIS method monitors seven inputs measured at the point of common coupling (PCC) such as root-mean square of voltage and current (RMSU and RMSI), total harmonic distortion of voltage and current (THDU and THDI), frequency, and active and reactive powers based on practical measurement in a real microgrid. The ANFIS method detects islanding condition by using its pattern recognition capability and nonlinear mapping of relation between input signals.

A novel hybrid islanding detection method for grid-connected microgrids in the case of multiple inverter-based distributed generators was proposed in [11]. This method was based on the slope of linear reactive power disturbance (RPD) and four passive criteria, namely, voltage variation, voltage unbalance, rate of change of frequency and correlation factor between the RPD and frequency variation. Islanding is detected when the frequency exceeds its thresholds. However, the cancellation issue in the case of multiple inverter-based distributed generators was ignored in this study.

In [12], a new method used the transient response of voltage waveform to detect the islanding condition. This method was based on two new criteria, namely, the peak of the transient index value (TIV) and the positive sequence superimposed phase angle at the point of common coupling.

In [13], a method based on Kalman filter (KF) was proposed to extract and filter the harmonic components of the measured voltage signal at distributed generation terminals. The islanding detected by the selected harmonic distortion (SHD) was calculated by the KF after the different changes in the power system. This changes was detected by a residual signal.

An unintentional islanding detection method based on the combination of the main criterion of switch status and the auxiliary criterion based on the no-current and difference-voltage criteria was proposed for "hand-in-hand" DC distribution network in [14]. However, the power supply reliability of the DC distribution network must be considered.

A novel islanding detection approach based on the relationship between the angular frequency and the derivative of the equivalent resistance observed from the small-scale synchronous generators in microgrids, including small-scale synchronous generators, was proposed in [15]. The derivative of the equivalent resistance observed from the small-scale synchronous generators approaches to zero when the islanding occurs in small-scale synchronous generator side.

In [16], we proposed an active islanding detection method by injecting perturbation signal to make the power imbalance and voltage fluctuation exceed the thresholds faster and easier in the islanding condition. This method can rapidly and efficiently detect the islanding condition. Howerver, the cancellation problem when using this method in multi-PV operation has not been considered.

Many converters are connected in parallel because of their capacity limitation in a large-scale solar system. Therefore, the cancellation issue can occur in a multi-PV operation.

On basis of the literature, no research has been conducted on the cancellation issue in multi-DG operation when using active islanding detection method. In this study, the cancellation issue when using the active Islanding Detection Method (IDM) in [16] is analyzed. Moreover, the solution improves the islanding detection method in [16] to eliminate the injected signal cancellation is proposed and verified by Matlab/Simulink. The system in this paper is the DC grid connected-photovoltaic system. The rest of this paper is organized as follows: The system description and cancellation problem are presented in Section 2. The simulation results of the cancellation problem is analyzed in Section 3. The solution for this problem is shown in Section 4. The conclusions are presented in Section 5.

2. System Description and Cancellation Problem

The injected signal cancellation in the DC grid-connected photovoltaic system is analyzed in this section.

2.1. System Description

As mentioned in the previous chapter, the cancellation signal occurs in multi-PV operation. To simplify, an operation system with two PVs is used for analysis, as shown in Figure 1.

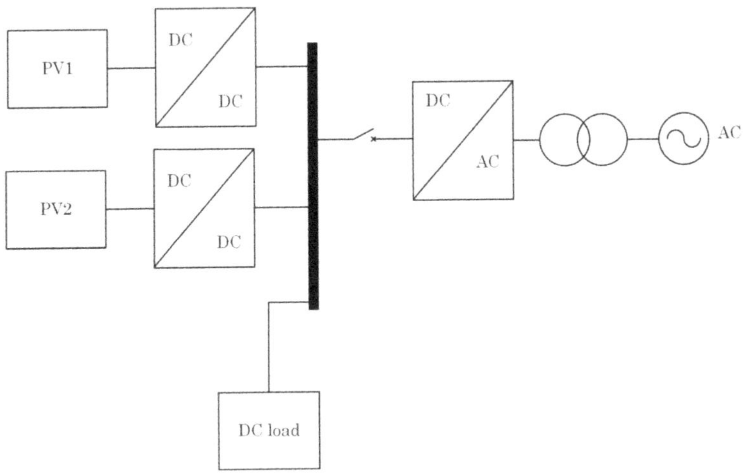

Figure 1. Cancellation testing system.

Figure 2 shows the diagram of the system under analysis. This system consists of a 100 kW PV array, output capacitor C, DC/DC converter with maximum power point tracking (MPPT), and islanding detection program. The DC load is shown as an equivalent resistance and the DC bus is modeled as a constant voltage source. The specifications of the PV module are listed in Table 1. The inverter is used to connect the AC and DC grids. A VSC controller is modeled as a voltage source converter (VSC-detail model) to regulate the DC bus voltage (keep stable at 500 V).

Table 1. Specifications of SunPower SPR-305E-WHT-D (SunPower, San Jose, CA, USA) PV module.

Information	Value
Open circuit voltage	64.2 V
Short circuit current	5.96 A
Voltage at maximum power point	54.7 V
Current at maximum power point	5.58 A
Maximum power	305.226 W
Parallel strings	66
Series-connected modules per string	5

The sampling time is $T_s = 5 \times 10^{-5}$. The simulation time is 2 s and the simulation is performed with the two following steps:

- Step 1: Start the injection of perturbation signal at 0.4 s after system startup.
- Step 2: The islanding event is activated at 1.2 s after system startup by disconnecting the AC grid.

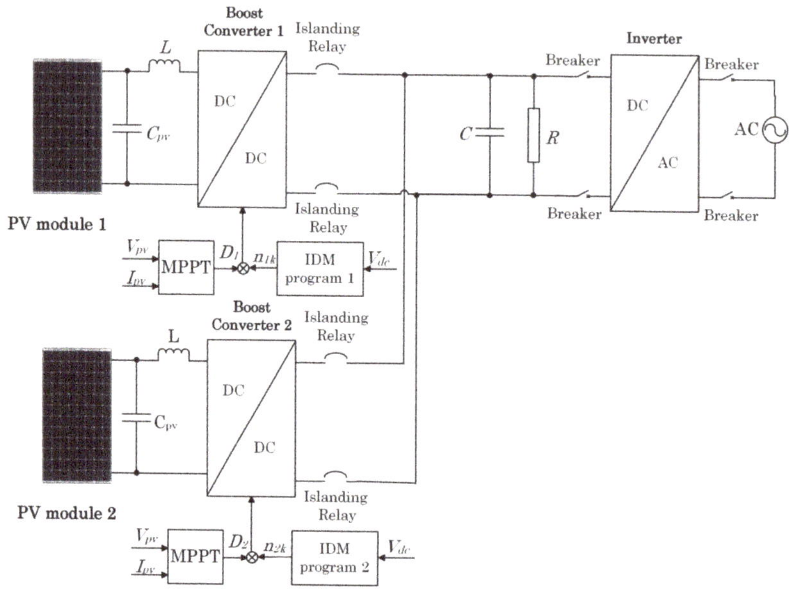

Figure 2. System model under analysis.

2.2. Injected Signal Cancellation

By using the proposed IDM in [16] in the multi-PV system, cancellation can occur when the perturbation signals of PV1 and PV2 are in the opposite positions, as shown in Figure 3.

- The perturbation signal factors at normal condition in normal case are n = 0.9–1.1.
- The perturbation signal factors at normal condition in the cancellation case are n = 1.1–0.9 for PV1 and n = 0.9–1.1 for PV2.

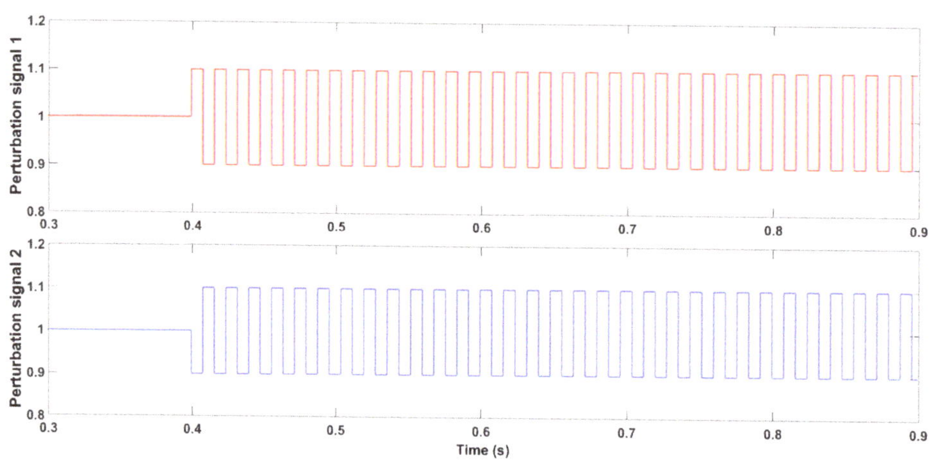

Figure 3. Perturbation signals in cancellation signal scenario.

Figure 4 describes in detail on how the injected signal cancellation can occur.

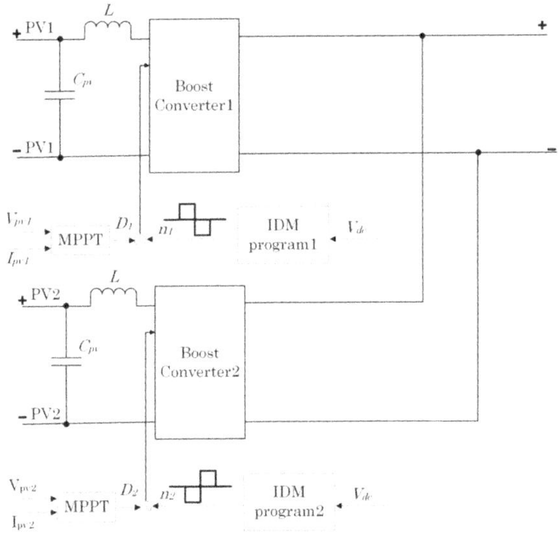

Figure 4. Controller perturbation signal circuit (V_{dc} is the measurement DC bus voltage; V_{pv1} and V_{pv2} are the PV1 and PV2 voltages, respectively; I_{pv1} and I_{pv2} are the PV1 and PV2 currents, respectively; n_1 and n_2 are the perturbation factors of IDM programs 1 and 2, respectively; D_1 and D_2 are the duty cycles of boost converter 1 and 2, respectively).

The cancellation problem is explained below:

Based on the equation in [16], after islanding, the DC-link voltage is shown in Equation (1):

$$v_{dc}(t) = [v(0) - N_k R i_{pv}(t)]e^{-\frac{1}{RC}t} + N_k R i_{pv}(t) \tag{1}$$

and

$$N_k = 1 - n_k D \tag{2}$$

where:

- $v_{dc}(t)$ is the DC-link voltage at time t.
- $v(0)$ is the initial DC-link voltage.
- $i_{pv}(t)$ is the photovoltaic current at time t.
- N_k is the output DC/DC converter current perturbation signal.
- n_k is the perturbation factor generated by the islanding program.
- D is the duty cycle.
- R is the DC load (resistor).
- C is the capacitor.

The DC-link voltage at perturbation N_1 from PV1 is as follows:

$$v_{dc1}(t) = N_1 R i_{pv}(t) + [v(0) - N_1 R i_{pv}(t)]e^{-\frac{1}{RC}t} \tag{3}$$

where: $v_{dc1}(t)$ is the DC-link voltage at time t when perturbation signal 1 is injected and, N_1 is the perturbation signal of PV1.

The DC-link voltage at perturbation N_2 from PV2 is shown in Equation (4):

$$v_{dc2}(t) = N_2 R i_{pv}(t) + [v(0) - N_2 R i_{pv}(t)]e^{-\frac{1}{RC}t} \tag{4}$$

where: $v_{dc2}(t)$ is the DC-link voltage at time t when perturbation signal 2 is injected.

The DC-link voltage without injecting the perturbation signal is as follows:

$$v_{dc0}(t) = N_0 R i_{pv}(t) + [v(0) - N_0 R i_{pv}(t)]e^{-\frac{1}{RC}t} \quad (5)$$

where: $v_{dc0}(t)$ is the DC-link voltage without perturbation signal at time t after islanding occurs.

The fluctuation of DC-link voltage caused by the perturbation signal from PV1 is shown in Equation (6):

$$v_{dc1}(t) - v_{dc0}(t) = \{N_1 R i_{pv}(t) + [v(0) - N_1 R i_{pv}(t)]e^{-\frac{1}{RC}t}\} - \{N_0 R i_{pv}(t) + [v(0) - N_0 R i_{pv}(t)]e^{-\frac{1}{RC}t}\} \quad (6)$$

The Equation (6) is simplified

$$v_{dc1}(t) - v_{dc0}(t) = (N_1 - N_0) R i_{pv}(t)(1 - e^{-\frac{1}{RC}t}) \quad (7)$$

Equation (7) becomes

$$v_{dc1}(t) - v_{dc0}(t) = (n_0 - n_1) D R i_{pv}(t)(1 - e^{-\frac{1}{RC}t}) \quad (8)$$

or

$$\Delta v_{1dc}(t) = (n_0 - n_1) D R i_{pv}(t)(1 - e^{-\frac{1}{RC}t}) \quad (9)$$

The fluctuation of DC-link voltage caused by the perturbation signal from PV2 is shown in Equation (10):

$$v_{dc2}(t) - v_{dc0}(t) = \{N_2 R i_{pv}(t) + [v(0) - N_2 R i_{pv}(t)]e^{-\frac{1}{RC}t}\} - \{N_0 R i_{pv}(t) + [v(0) - N_0 R i_{pv}(t)]e^{-\frac{1}{RC}t}\} \quad (10)$$

The Equation (10) is simplified

$$v_{dc2}(t) - v_{dc0}(t) = (N_2 - N_0) R i_{pv}(t)(1 - e^{-\frac{1}{RC}t}) \quad (11)$$

Equation (11) becomes

$$v_{dc2}(t) - v_{dc0}(t) = (n_0 - n_2) D R i_{pv}(t)(1 - e^{-\frac{1}{RC}t}) \quad (12)$$

or

$$\Delta v_{2dc}(t) = (n_0 - n_2) D R i_{pv}(t)(1 - e^{-\frac{1}{RC}t}) \quad (13)$$

Based on Equations (9) and (13), the injected signal cancellation occurs when:

$$\Delta v_{1dc}(t) + \Delta v_{2dc}(t) = 0 \quad (14)$$

$$2n_0 = n_1 + n_2 \quad (15)$$

If n_0 = 1 (without perturbation signal), n_1 = 1.1, and n_2 = 0.9, then Equation (15) is satisfied. Consequently, the fluctuation of DC-link voltage is cancelled because the perturbation signals from PV1 and PV2 have the same value but opposite directions. Thus, the DC-link voltage fluctuations have been cancelled.

The simulation result verifies the problem.

3. Simulation Testing Scenarios and Results

The system is tested under the balance case between the DC load and PV in Table 2, and the procedure is as follows:

- The perturbation signal is injected at t = 0.4 s after the system startup.
- The islanding condition occurs at t = 1.2 s after the system startup.

The model parameters are as follows:

- The maximum power (each PV) is P_{max} = 50 kW.
- Switching frequency of the converter is f_s = 10 kHz.
- The DC bus voltage is V_{dc} = 500 V.
- The under voltage threshold is V_{dcmin} = 450 V.
- The islanding threshold is 0.2346 [16].
- The perturbation duration is T = 8 ms.
- The perturbation signal factors at normal condition in the normal case are n = 0.9–1.1.
- The perturbation signal factors at normal condition in the cancellation case are n = 1.1–0.9 for PV1 and n = 0.9–1.1 for PV2.

Table 2. Normal and cancellation scenarios.

Cases	PV1 (kW)	PV2 (kW)	Total PV (kW)	Load (kW)	Vdc(V)
Normal	50	50	100	100	500
Cancellation	50	50	100	100	500

For the first test, the IDM will test two scenarios: normal and cancellation. The results are shown in the Figures 5 and 6.

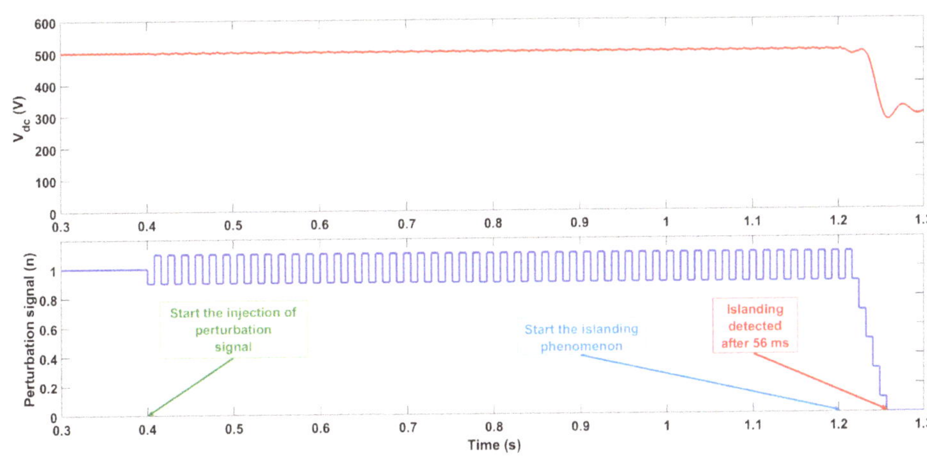

Figure 5. Islanding detection result by using the IDM in [16] in normal scenario.

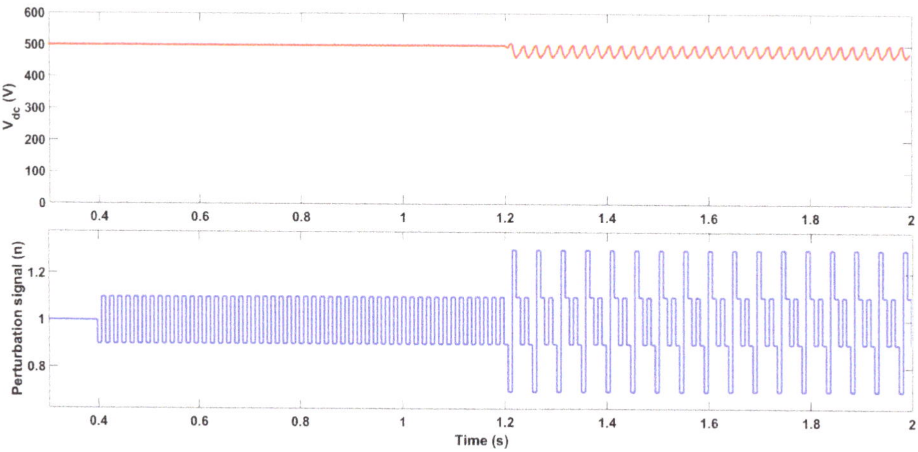

Figure 6. Islanding detection result by using the IDM in [16] in cancellation scenario.

The result in the normal scenario is shown in Figure 5, and the cancellation scenario result is shown in Figure 6. Figure 5 shows that the islanding condition is detected after 56(ms). The IDM cannot detect the islanding condition when the cancellation problem occurs. In closing, injecting two perturbation signals in the opposite directions can make the cancellation occur completely.

The impact of the injected signal cancellation on the islanding detection method is significant based on the results. The islanding detection method cannot detect the islanding phenomenon in the cancellation scenario.

4. Improved Islanding Detection Method Based on the Proposed Solution

According to the previous section, the cancellation problem occurs by using two perturbations in the opposite directions. The main meaning of the solution is how to prevent the occurrence of cancellation.

First, the beginning of the proposed IDM uses the perturbation signal in the range n = 0–1.1. As the range of perturbation signal includes values over and below the normal value (n = 1), the injected signal cancellation caused by the perturbation signals in opposite directions can occur.

Second, to prevent the occurrence of cancellation, the perturbation signal value is changed so that the opposite signals cannot occur.

Therefore, the solution for the cancellation problem is to change the perturbation signal values in the range n = 0–1.1 to n = 0–1, and in the normal condition from n = 0.9–1.1 to n = 0.9–1. By changing this value, the injected signal cancellation is eliminated. Similar to the Equations (3)–(15), the solution is explained below:

The injected signal cancellation occurs when:

$$2n_0 = n_1 + n_2 \tag{16}$$

By using the new perturbation signal (n_0 = 1, n_1 = 1, and n_2 = 0.9), Equation (16) becomes

$$2 \neq 1.9 \tag{17}$$

Based on Equation (17), the fluctuation of DC-link voltage caused by the perturbation signals from PV1 and PV2 cannot cancel each other. Thus, the proposed solution can solve the injected signal cancellation.

By generalizing this solution with n_k perturbation signals ($k = 1 \div \infty$), the injected signal cancellation occurs when:

$$\Delta v_{1dc}(t) + \Delta v_{2dc}(t) + \ldots + \Delta v_{kdc}(t) = 0 \tag{18}$$

or

$$n_0 - n_1 + n_0 - n_2 + \ldots + n_0 - n_k = 0 \tag{19}$$

Finally,

$$kn_0 = \sum_{k=1}^{\infty} n_k \tag{20}$$

As n_0 = 1 (without perturbation signal) and $n_k \leq 1$, Equation (20) is correct when all perturbation signal values are 1 (without perturbation signal case).

With other values of the perturbation signals (with perturbation signal), Equation (20) becomes

$$kn_0 > \sum_{k=1}^{\infty} n_k \tag{21}$$

For this reason, the injected signal cancellation is eliminated by using the proposed solution.

The flowchart in Figure 7 explains the detailed procedure of the proposed solution. The solution is verified by a step-by-step simulation.

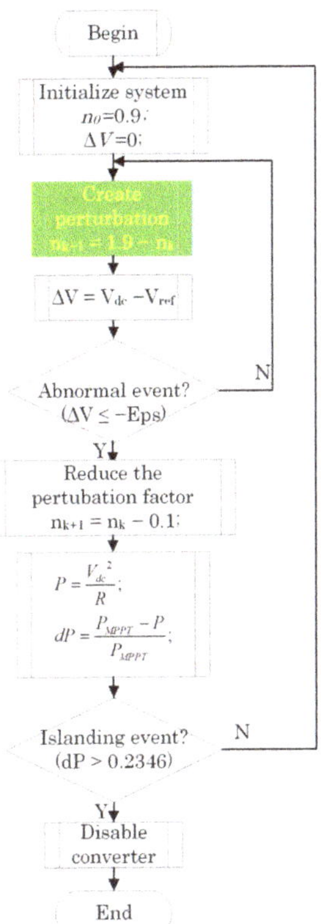

Figure 7. Flowchart of the improved IDM program (n_0 is the perturbation factor at the beginning, ΔV is the voltage deviation, Eps is the abnormal event value, dP is the rate of change of output power).

The testing scenarios in Table 3 are the worst case (the power of DGs and the load are balance), and the procedure is similar to the previous testing procedure.

Table 3. Multi-PV operation scenario.

Cases	PV1(kW)	PV2(kW)	PV3(kW)	PV4(kW)	Total PV(kW)	Load(kW)	Vdc(V)
2-PV	50	50	0	0	100	100	500
3-PV	50	50	50	0	150	150	500
4-PV	50	50	50	50	200	200	500

The model parameters are almost similar to the previous testing, only the new perturbation factors are different. The perturbation factors at normal condition in cancellation case are n = 1–0.9 for PV1 and n = 0.9–1 for PV2.

In addition, the eight-PV scenario also tests to verify the effect of the improved IDM. The parameters of the eight-PV scenario are as below:

- Each PV module power is 50 kW.
- Total PV power is 400 kW.
- Load power is 400 kW.
- DC bus voltage is 500 V.

As shown in the results in Figure 8, the injected signal cancellation occurs in two-PV scenario. With the improved islanding detection method based on the proposed solution, the islanding condition is detected after 40 ms when the islanding occurs. The solution is effective in two-PV scenario.

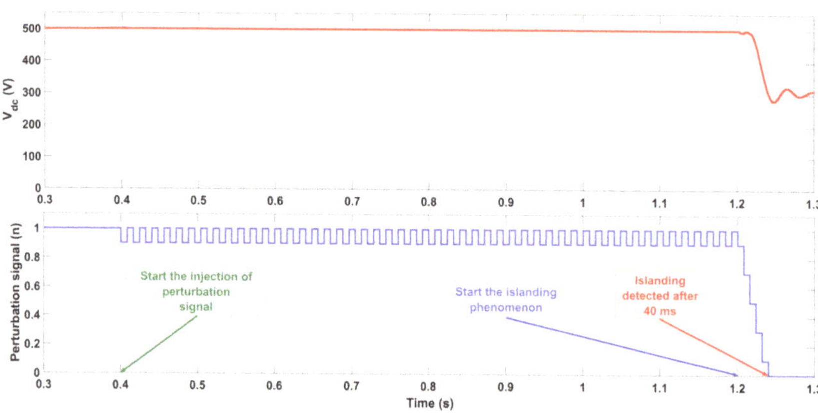

Figure 8. Solution result in two-PV scenario.

The results of three-PV scenario are shown in Figure 9 and 10. In Figure 9, the islanding detection method can not detect the islanding phenomenon when injected signal cancellation occurs.

The results in Figure 10 show that the islanding condition is detected after 64 ms when the islanding occurs by applying the proposed solution to the improved islanding detection method in the injected signal cancellation case.

The four-PV scenario has the similar results, the islanding condition can be detected after 48 ms when the islanding occurs by using the improved islanding detection method, but it cannot be detected by using the original one. The results are shown in Figures 11 and 12.

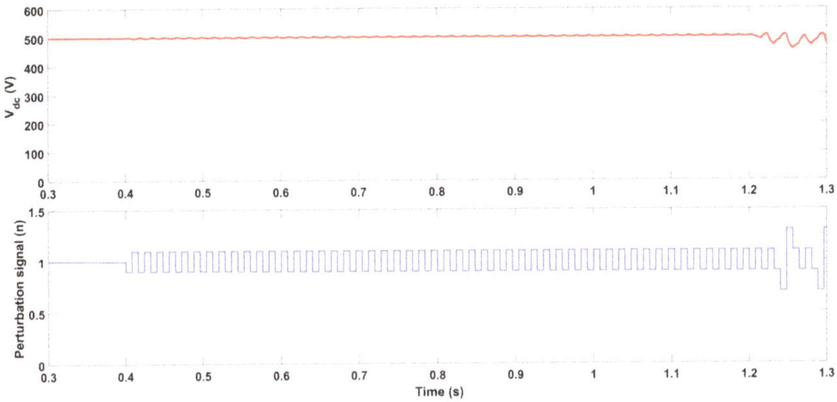

Figure 9. Cancellation result in three-PV scenario.

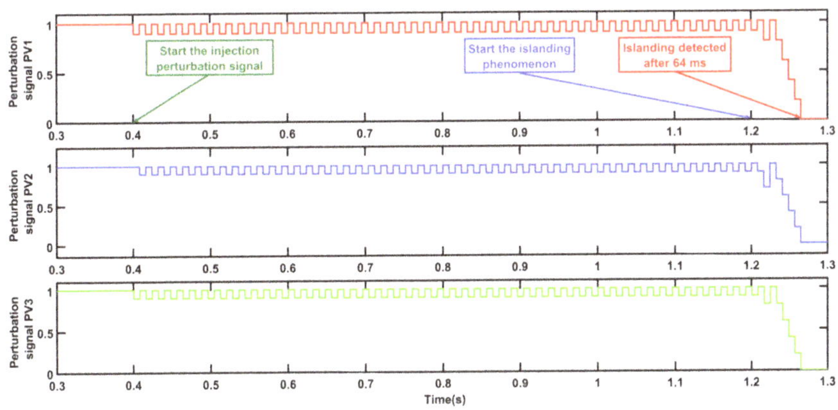

Figure 10. Solution result in three-PV scenario.

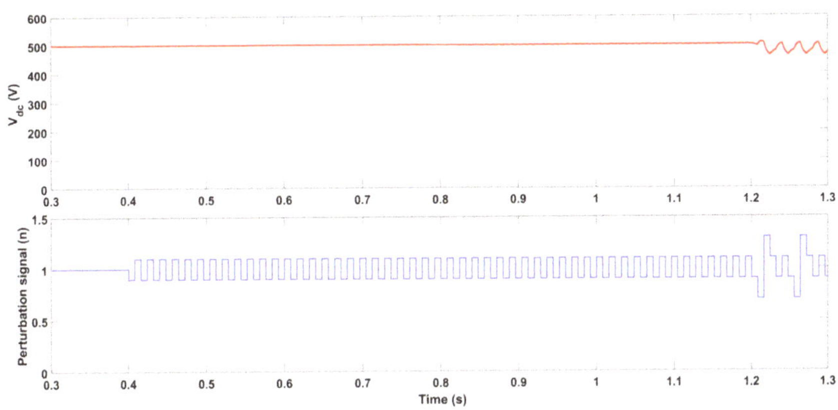

Figure 11. Cancellation result in four-PV scenario.

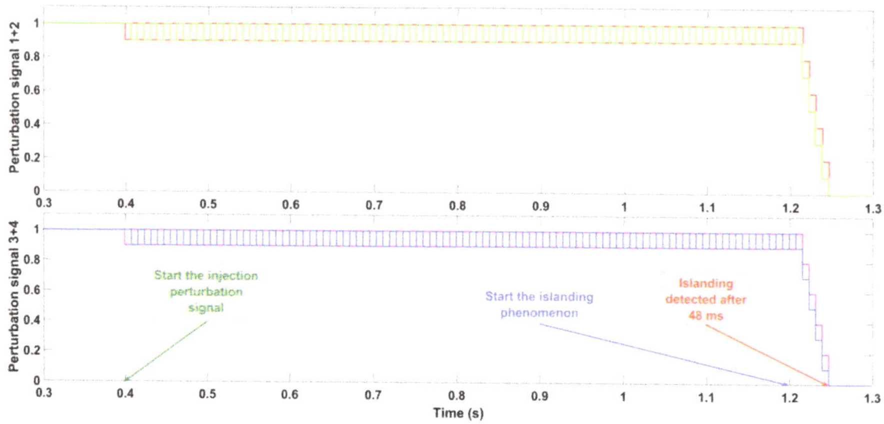

Figure 12. Solution result in four-PV scenario.

The eight-PV scenario result is shown in Figure 13. The improved islanding detection method by applying the proposed solution can detect the islanding condition after 48 ms.

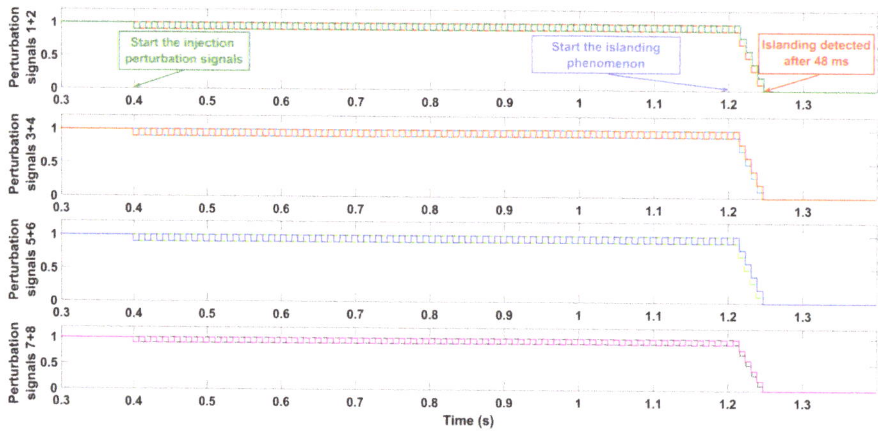

Figure 13. Solution result in eight-PV scenario.

Based on the simulation results and mathematical explanations, the proposed solution eliminates the injected signal cancellation by using the improved islanding detection method based on the proposed solution in two-PV, three-PV, four-PV, and eight-PV scenarios.

Moreover, the injected signal cancellation is eliminated not only in the specified cases but also in the general case by using the improved islanding detection method. This result shows the achievement and efficiency of the proposed solution.

5. Conclusions

In this study, the islanding detection method based on perturbation signal injection and the rate of change of output power in a DC grid-connected photovoltaic system is investigated.

The impact of injected signal cancellation on the islanding detection method in [16] is significant. The islanding detection method in [16] cannot detect the islanding phenomenon but works well without injected signal cancellation (detection time is 56 ms).

The solution of using the new perturbation factors (n = 0.9–1 in normal condition) is proposed. With this solution, the improved islanding detection method can detect the islanding phenomenon after 40 ms (two-PV scenario), 64 ms (three-PV scenario), and 48 ms (four-PV and eight-PV scenarios) when the injected signal cancellation appears.

Through the mathematical explanations and simulation results, the injected signal cancellation is eliminated. Moreover, the improved islanding detection method by applying the proposed solution is correct not only in specified cases but also in the general case.

Author Contributions: T.S.T. and D.T.N. extensively reviewed the literature and drafted the manuscript and G.F. provided feedback and critical comments for inclusions and finally helped in revising the manuscript to improve the quality of paper. All authors provided critical feedback and helped shape the research, analysis and manuscript. Besides, all authors discussed the results and contributed to the final manuscript.

Funding: This research received no external funding.

Conflicts of Interest: The authors declare no conflict of interest.

References

1. Elsayed, A.T.; Mohamed, A.A.; Mohammed, O.A. Review DC microgrids and distribution systems: An overview. *Electr. Power Syst. Res.* **2015**, *119*, 407–417. [CrossRef]
2. Hammerstrom, D.J. AC Versus DC Distribution Systems: Did We Get it Right?. In Proceedings of the IEEE Power Engineering Society General Meeting, Tampa, FL, USA, 24–28 June 2007; pp. 1–5.
3. Seo, G.-S.; Baek, J.; Choi, K.; Bae, H.; Cho, B. Modeling and analysis of DC distribution systems. In Proceedings of the IEEE 8th International Conference on Power Electronics and ECCE Asia, Jeju, Korea, 30 May–3 June 2011; pp. 223–227.
4. Starke, M.R.; Tolbert, L.M.; Ozpineci, B. AC vs DC distribution: A loss comparison. In Proceedings of the IEEE PES Transmission and Distribution Conference and Exposition, Chicago, IL, USA, 21–24 April 2008; pp. 1–7.
5. Sannino, A.; Postiglione, G.; Bollen, M. Feasibility of a DC Network for Commercial Facilities. *IEEE Trans. Ind. Appl.* **2003**, *39*, 1499–1507. [CrossRef]
6. IEEE. *IEEE Recommended Practice for Utility Interface of Photovoltaic (PV) Systems*; IEEE Std 929-2000; IEEE: Piscataway, NJ, USA, 2000; doi:10.1109/IEEESTD.2000.91304. [CrossRef]
7. Zamani, R.; Hamedani-Golshan, M.-E.; Alhelou, H.H.; Siano, P.; Pota, H.R. Islanding Detection of Synchronous Distributed Generator Based on the Active and Reactive Power Control Loops. *Energies* **2018**, *11*, 2819. [CrossRef]
8. Ghalavand, F.; Alizade, B.A.M.; Gaber, H.; Karimipour, H. Microgrid Islanding Detection Based on Mathematical Morphology. *Energies* **2018**, *11*, 2696. [CrossRef]
9. Di Fazio, A.R.; Russo, M.; Valeri, S. A New Protection System for Islanding Detection in LV Distribution Systems. *Energies* **2015**, *8*, 3775–3793. [CrossRef]
10. Mlakić, D.; Baghaee, H.R.; Nikolovski, S. A Novel ANFIS-Based Islanding Detection for Inverter-Interfaced Microgrids. *IEEE Trans. Smart Grid* **2019**, *10*, 4411–4424, doi:10.1109/TSG.2018.2859360. [CrossRef]
11. Chen, X.; Li, Y.; Crossley, P. A Novel Hybrid Islanding Detection Method for Grid-Connected Microgrids with Multiple Inverter-Based Distributed Generators Based on Adaptive Reactive Power Disturbance and Passive Criteria. *IEEE Trans. Power Electron.* **2019**, *34*, 9342–9356, doi:10.1109/TPEL.2018.2886930. [CrossRef]
12. Nale, R.; Biswal, M.; Kishor, N. A Transient Component Based Approach for Islanding Detection in Distributed Generation. *IEEE Trans. Sustain. Energy* **2019**, *10*, 1129–1138. [CrossRef]
13. Haider, R.; Kim, C.H.; Ghanbari, T.; Bukhari, S.B.A. Harmonic-signature-based islanding detection in grid-connected distributed generation systems using Kalman filter. *IET Renew. Power Gener.* **2018**, *12*, 1813–1822. [CrossRef]
14. Song, W.; Chen, Y.; Wen, A.; Zhang, Y.; Wei, C. Detection and switching control scheme of unintentional islanding for 'hand-in-hand' DC distribution network. *IET Gener. Transm. Distrib.* **2019**, *13*, 1414–1422. [CrossRef]

15. Bekhradian, R.; Davarpanah, M.; Sanaye-Pasand, M. Novel Approach for Secure Islanding Detection in Synchronous Generator Based Microgrids. *IEEE Trans. Power Deliv.* **2019**, *34*, 457–466. [CrossRef]
16. Son, T.T.; Tuyen, N.D.; FUJITA, G. Islanding Detection Method Based on Injecting Perturbation Signal and Rate of Change of Output Power in DC Grid-Connected Photovoltaic System. *Energies* **2018**, *11*, 1313. [CrossRef]

applied sciences

MDPI

Article

Insulation Monitoring Method for DC Systems with Ground Capacitance in Electric Vehicles

Jifei Du [1], Trillion Q. Zheng [1], Yian Yan [2,*], Hongyan Zhao [1], Yangbin Zeng [1] and Hong Li [1]

1 School of Electrical Engineering, Beijing Jiaotong University, No. 3 Shang Yuan Cun, Haidian District, Beijing 100044, China

2 National Active Distribution Network Technology Research Center (NANTEC), Beijing Jiaotong University, Beijing 100044, China

* Correspondence: yanyian@bjtu.edu.cn; Tel.: +86-10-5168-4056; Fax: +86-10-5168-3907

check for updates

Received: 21 May 2019; Accepted: 25 June 2019; Published: 27 June 2019

Featured Application: The insulation monitoring generally applied to DC charging piles, battery management systems, or high voltage distribution system for electric vehicles. It is effective on improving the safety performance of electric vehicles.

Abstract: Owing to the influence of ground capacitance in electric vehicles, in the traditional unbalanced electric bridge DC insulation monitoring (DC-IM) method, the voltage of positive and negative electric bridges changes slowly. To calculate the insulation resistances, sampling should be conducted once the voltage of the bridge becomes stable, that will inevitably extend the monitoring cycle. To reduce the monitoring cycle, this study proposes a three-point climbing algorithm, namely, three-bridge voltage sampling with equal sampling intervals, to predict the evolution of the bridge voltage curve. However, due to the existence of sampling errors, the insulation resistances calculated by sampling values will deviate from the actual values. Then, this article also proposes the filter and correction methods of three sampled voltages to improve monitoring accuracy. Through experimental data, the influences of different parameters on the results are verified, and comparisons with the traditional method are shown in the back. The conclusion is that compared with the traditional method, the proposed method can monitor insulation resistance more quickly and ensure fixed monitoring cycles under different ground capacitance values and keep the similar monitoring accuracy.

Keywords: unbalanced electric bridge; DC insulation monitoring; ground capacitance; three-point climbing algorithm

1. Introduction

With the increasing popularity of electric vehicles (EVs), strict requirements are being established for the driving and charging safety of EVs. The insulation of EVs decreases due to rain, dampness, collision, and other reasons arising from the long-term exposure of EVs and charging equipment to the outdoor environment [1,2]. The DC system of EVs is connected to numerous power electronic devices, including the motor converter, battery charger, air conditioner, and DC–DC converter [3], and the overall connections may form a DC power microgrid system when EV is charging [4,5]. Insulation failure of any equipment affects the safety of the entire system. When the insulation resistance of the system decreases to below a threshold value, the vehicle sends warning signals. If the situation is serious, then the high-voltage system must be cut off and stopped for troubleshooting [6,7]. The DC insulation monitoring (DC-IM) function is thus required before charging by the DC charging pile and during the process of driving the EVs [8]. Various insulation-monitoring devices and embedded circuits have been designed and installed in DC charging piles, battery packs, high-voltage distribution boxes, and other equipment or embedded in the battery management system of EVs. DC-IM methods include balanced electric bridge [9], unbalanced electric bridge [10–14], high-voltage injection [15], differential amplification [16], and low-frequency small-signal injection [17,18]. The unbalanced electric bridge method can synchronously monitor positive and negative insulation resistances, has a low cost, and is easy to realize; thus, it has been widely used in EVs and charging piles.

Because the DC system of EVs connects various power electronic devices that contain many Y capacitors and parasitic capacitors, which make up the large ground capacitance (GC) of the system, GC, an unknown system parameter, seriously affects the monitoring accuracy and speed of DC-IM. Therefore, various solutions have been suggested in the literature. In [19,20], wavelet-transform and chaos theory detection methods were proposed to deal with interference signals. However, these methods are more suitable for a multi-branch DC system with the small-signal injection method than for systems with a large GC. The method based on the Kalman filter and Lyapunov equation proposed in [18] and [21] needs to be recursive step by step. Thus, obtaining the result takes a long time. The traditional sampling and comparison method is frequently used in current practical product applications. After initiating bridge conversion, sampling and calculation are performed only after GC is fully charged so that a stable voltage signal can be sampled. However, this method considerably slows down DC-IM and cannot meet the real-time requirements of EVs and the future development trend of EV safety.

This study proposes a method of unbalanced electric bridge DC-IM based on a three-point climbing algorithm. After switching the bridge, sampling is conducted for three times at equal intervals. The methods of filtering and automatic correction of sampling voltage are also proposed to reduce the result error caused by voltage ripple and sampling resolution. The calculation can predict the voltage value after the completion of GC charging. The method is simple and easy to implement. It does not need to wait for GC charging and multiple sampling, which can considerably increase the detection speed. Thus, the DC-IM period is fixed and unaffected by GC.

The rest of this paper is organized as follows. Section 2 analyzes the unbalanced electric bridge DC-IM with the existing GC. Section 3 proposes the novel method of the three-point climbing algorithm in order to avoid the impact on GC. Section 4 further optimizes the proposed method and describes the implementation method. In Section 5, The experimental data are exhibited to prove the theory. Finally, conclusions are included in Section 6. Some symbols used in the operation optimization are shown in Table 1.

Table 1. List of some symbols used in the operation optimization.

Symbol	Explanation
DC-IM	DC insulation monitoring
GC	Ground capacitance
R_{f1}	The negative insulation resistance
R_{f2}	The positive insulation resistance
R_a	The larger bridge resistance
R_b	The smaller bridge resistance
C_1	The capacitance value of the DC negative pole to the earth
C_2	The capacitance value of the DC positive pole to the earth
v_1	The voltage value of the DC negative pole to the earth
v_2	The voltage value of the DC positive pole to the earth
v_{11} and v_{12}	The sample value of v_1 in M_1 and M_2 phase respectively
v_{21} and v_{22}	The sample value of v_2 in M_1 and M_2 phase respectively
Δt	The time interval of sampling
E_1	$e^{-\frac{\Delta t}{\tau_1}}$ of M_1 phase
E_2	$e^{-\frac{\Delta t}{\tau_2}}$ of M_2 phase
$\hat{E}_1(k)$	The estimated value of the kth E_1
$\hat{E}_2(k)$	The estimated value of the kth E_2
$\widetilde{v}$	Estimated voltage value of the half-bridge voltage
$\hat{v}$	Correction voltage value of the half-bridge voltage

2. Traditional Unbalanced Electric Bridge DC-IM Method and Analysis of the Influence on GC

The unbalanced electric bridge DC-IM topology circuit is shown in Figure 1, where v_{dc} is the DC voltage; R_{f2} and R_{f1} are the positive and negative insulation resistances, respectively; R_a and R_b are the bridge resistances, $R_a = R_1$, $R_b = R_1||R_2$, so $R_a > R_b$. The unbalanced electric bridge method works in two phases, namely, M_1 and M_2. In the M_1 phase, Q_1 turn-on and Q_2 turn-off, the positive half-bridge resistance is R_a, the negative half-bridge resistance is R_b, the negative and positive half-bridge voltages are v_{11} and v_{12}, respectively, and the ground current is i_1, as shown in Figure 1a. In the M_2 phase, Q_1 turn-off and Q_2 turn-on, the positive half-bridge resistance is R_b, and the negative half-bridge resistance is R_a. The negative and positive half-bridge voltages are v_{21} and v_{22}, respectively, and the ground current is i_2, as shown in Figure 1b. The conventional DC-IM method can be expressed as Equation (1) by Kirchhoff's law.

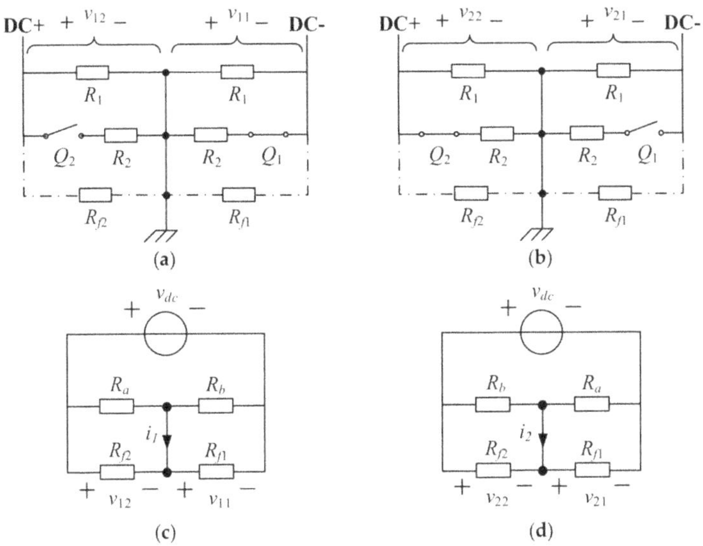

Figure 1. The circuit of unbalanced bridge. (**a**) The circuit of M_1 phase. (**b**) The circuit of M_2 phase. (**c**) The equivalent circuit of M_1 phase. (**d**) The equivalent circuit of M_2 phase.

$$\begin{cases} R_{f1} = \frac{v_{dc}(R_a-R_b)-(i_1-i_2)(R_aR_b)}{i_1R_b-i_2R_a} \\ R_{f2} = \frac{v_{dc}(R_a-R_b)-(i_1-i_2)(R_aR_b)}{i_1R_a-i_2R_b} \end{cases} \quad (1)$$

When the DC system has GC, the capacitance value of the DC negative pole to the earth is C_1, and the capacitance value of the DC positive pole to the earth is C_2. Thus, the circuits in Figure 1c,d can be changed to those shown in Figure 2a,b. To facilitate calculation and analysis, the equivalent resistance of the two working modes is assumed to be what is shown in Equation (2). Figure 2a,b can be simplified as Figure 2c,d.

$$\begin{cases} R_{11} = \frac{R_bR_{f1}}{R_b+R_{f1}}, R_{12} = \frac{R_aR_{f2}}{R_a+R_{f2}} \\ R_{21} = \frac{R_aR_{f1}}{R_a+R_{f1}}, R_{22} = \frac{R_bR_{f2}}{R_b+R_{f2}} \end{cases} \quad (2)$$

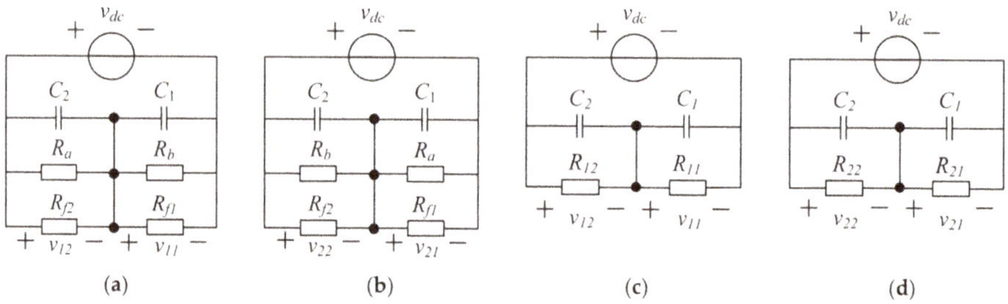

Figure 2. Equivalent circuit of unbalanced bridge with GC. (**a**) The circuit of M_1 phase. (**b**) The circuit of M_2 phase. (**c**) The equivalent circuit of M_1 phase. (**d**) The equivalent circuit of M_2 phase.

The following parameters are set.

$$\begin{cases} X_{11} = \frac{R_{11}}{R_{11}+R_{12}}, X_{12} = \frac{R_{12}}{R_{11}+R_{12}} \\ X_{21} = \frac{R_{21}}{R_{21}+R_{22}}, X_{22} = \frac{R_{22}}{R_{21}+R_{22}} \end{cases} \quad (3)$$

According to Figure 2, the time constants of the M_1 and M_2 phases are defined as

$$\tau_1 = \frac{(C_1 + C_2)R_{11}R_{12}}{(R_{11} + R_{12})} \text{ and } \tau_2 = \frac{(C_1 + C_2)R_{21}R_{22}}{(R_{21} + R_{22})} \quad (4)$$

When the two phases switch with each other, the charging process of GC belongs to the first-order circuit full response process, and the curvilinear function Equation (5) can be obtained, where v_{110}, v_{120}, v_{210}, and v_{220} are the initial voltage of the full response processes of v_{11}, v_{12}, v_{21}, and v_{22}, respectively.

$$\begin{cases} v_{11} = v_{dc}X_{11} + (v_{110} - v_{dc}X_{11})e^{-\frac{t}{\tau_1}} \\ v_{12} = v_{dc}X_{12} + (v_{120} - v_{dc}X_{12})e^{-\frac{t}{\tau_1}} \\ v_{21} = v_{dc}X_{21} + (v_{210} - v_{dc}X_{21})e^{-\frac{t}{\tau_2}} \\ v_{22} = v_{dc}X_{22} + (v_{220} - v_{dc}X_{22})e^{-\frac{t}{\tau_2}} \end{cases} \quad (5)$$

3. New Strategy to Avoid the Impact on GC: Three-Point Climbing Algorithm

The traditional sampling method when GC exists is shown in Figure 3. In the M_1 phase, v_1 decreases slowly, and v_2 increases slowly. Sampling continues for v_1 and v_2 until t_{11_n} and t_{12_n}, respectively; v_1 falls to stable value $v_{dc}X_{11}$, and v_2 rises to stable value $v_{dc}X_{12}$. After stabilization, the final v_{11} and v_{12} are obtained. The method switches to the M_2 phase until v_1 and v_2 reach stable values $v_{dc}X_{21}$ and $v_{dc}X_{22}$, respectively, and the controller obtains the final v_{21} and v_{22}. One sampling period T_C ends, and the insulation resistance values R_{f2} and R_{f1} are calculated by $v_{11} = v_{dc}X_{11}$, $v_{12} = v_{dc}X_{12}$, $v_{21} = v_{dc}X_{21}$, and $v_{22} = v_{dc}X_{22}$. The resistance values of R_{11}, R_{12}, R_{21}, and R_{22} are large, so the charging time of the capacitor is long. Sampling and calculation should be performed after GC charging is completed; hence, the measurement time of the traditional method is long and unable to meet the real-time requirements of EVs. To avoid the measurement overtime caused by GC, a new insulation resistance monitoring method, namely, three-point climbing algorithm, is proposed.

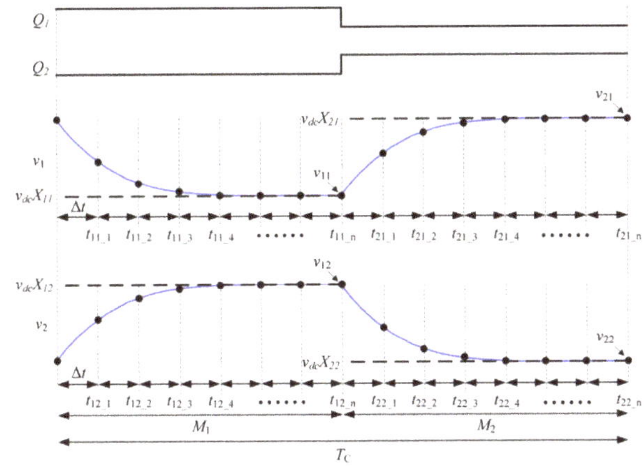

Figure 3. Traditional sampling method.

Figure 4 shows that each phase is sampled three times, and the sampling intervals are equal. With v_{11} as an example, the three sampling times are t_{11_1}, t_{11_2}, and t_{11_3}; the sampling voltage values are v_{11_1}, v_{11_2}, and v_{11_3}, respectively, and the time intervals are Δt. Similar definitions of v_{12}, v_{21}, and v_{22} are provided to facilitate the calculation, and E_1 and E_2 are used to present natural exponential function. The following equation is then obtained from (5).

$$\begin{cases} \frac{v_{11\_1}-v_{11\_2}}{v_{11\_2}-v_{11\_3}} = e^{-\frac{\Delta t}{\tau_1}} = E_1, & \frac{v_{12\_1}-v_{12\_2}}{v_{12\_2}-v_{12\_3}} = e^{-\frac{\Delta t}{\tau_1}} = E_1 \\ \frac{v_{21\_1}-v_{21\_2}}{v_{21\_2}-v_{21\_3}} = e^{-\frac{\Delta t}{\tau_2}} = E_2, & \frac{v_{22\_1}-v_{22\_2}}{v_{22\_2}-v_{22\_3}} = e^{-\frac{\Delta t}{\tau_2}} = E_2 \end{cases} \tag{6}$$

t_{11_1}, t_{12_1}, t_{21_1}, and t_{22_1} are set as initial times for the first-order circuit full response curve, and the curve Equation (5) is converted into Equation (7).

$$\begin{cases} v_{11\_2} = v_{dc}X_{11} + (v_{11\_1} - v_{dc}X_{11})E_1 \\ v_{12\_2} = v_{dc}X_{12} + (v_{12\_1} - v_{dc}X_{12})E_1 \\ v_{21\_2} = v_{dc}X_{21} + (v_{21\_1} - v_{dc}X_{21})E_2 \\ v_{22\_2} = v_{dc}X_{22} + (v_{22\_1} - v_{dc}X_{22})E_2 \end{cases} \tag{7}$$

Equation (7) is converted into Equation (8).

$$\begin{cases} X_{11} = \frac{v_{11\_2}-v_{11\_1}E_1}{v_{dc}(1-E_1)}, & X_{12} = \frac{v_{12\_2}-v_{12\_1}E_1}{v_{dc}(1-E_1)} \\ X_{21} = \frac{v_{21\_2}-v_{12\_1}E_2}{v_{dc}(1-E_2)}, & X_{22} = \frac{v_{22\_2}-v_{22\_1}E_2}{v_{dc}(1-E_2)} \end{cases} \tag{8}$$

X_{11}, X_{12}, X_{21}, and X_{22} in Equation (8) are known values calculated by sampling voltage. Equation (9) can be obtained from Equations (2), (3), and (8), and insulation resistance values R_{f1} and R_{f2} can be solved.

$$\begin{cases} R_{f1} = \frac{R_a R_b}{\frac{R_a-R_b}{X_{12}-X_{22}}X_{12}-R_a}, R_{f2} = \frac{R_a R_b}{\frac{R_b-R_a}{X_{11}-X_{21}}X_{11}-R_b} \\ R_{f1} = \frac{R_a R_b}{\frac{R_a-R_b}{X_{12}-X_{22}}X_{22}-R_b}, R_{f2} = \frac{R_a R_b}{\frac{R_b-R_a}{X_{11}-X_{21}}X_{21}-R_a} \end{cases} \tag{9}$$

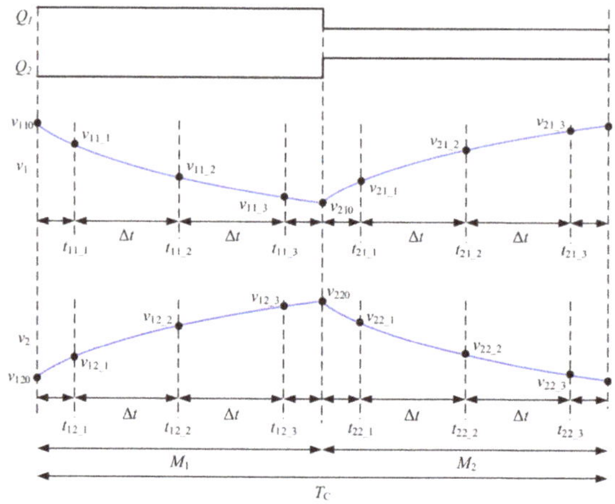

Figure 4. Proposed three-point climbing algorithm.

4. Implementation Method for Improving Accuracy

4.1. Error Analysis

In practical applications, sample resolution and voltage ripple cause sampling errors, which affect the calculated result of insulation resistance. With the v_{11} of the M_1 phase as an example, Δv_{11_1}, Δv_{11_2}, and Δv_{11_3} are the sampling errors of v_{11_1}, v_{11_2}, and v_{11_3}, respectively. E_{1R} is the actual value of E_1. E_{1C} is the measured value of E_1. Considering the sampling error, the expression of E_{1R} and E_{1C} according to Equation (6) is shown as

$$\begin{cases} E_{1R} = \frac{v_{11\_2} - v_{11\_3}}{v_{11\_1} - v_{11\_2}} \\ E_{1C} = \frac{(v_{11\_2} + \Delta v_{11\_2}) - (v_{11\_3} + \Delta v_{11\_3})}{(v_{11\_1} + \Delta v_{11\_1}) - (v_{11\_2} + \Delta v_{11\_2})} \end{cases} \tag{10}$$

ΔE_1 is the error of E_1 and is defined as follows:

$$\Delta E_1 = E_{1C} - E_{1R} \tag{11}$$

By substituting Equation (11) into Equation (10), ΔE_1 can be rewritten as

$$\Delta E_1 = \frac{(\Delta v_{11\_2} - \Delta v_{11\_3}) - E_{1R}(\Delta v_{11\_1} - \Delta v_{11\_2})}{(v_{11\_1} - v_{11\_2}) + (\Delta v_{11\_1} - \Delta v_{11\_2})} \tag{12}$$

where Δv_{11_1}, Δv_{11_2}, and Δv_{11_3} are uncontrollable components and E_{1R} is a fixed value. $(v_{11\_1} - v_{11\_2})$ is inversely proportional to ΔE_1. Similarly, v_{12}, v_{21}, and v_{22} can result in the same conclusion. In the application, the larger Δt_{11}, Δt_{12}, Δt_{21}, and Δt_{22} are, the smaller the result error is. The larger the difference between the R_a and R_b is, the smaller the result error is. The higher the v_{dc} is, the smaller the result error is.

4.2. Selection of Calculation Method

In the climbing stage, the smaller GC is, the larger the difference is in the three-point sampling and the smaller the result error is. If GC is too small, the sampling may reach the stable stage after climbing, the difference between the three sampled voltages will be too small, and the resulting error will increase. The accuracy is the highest only when the three sampled voltages are in the climbing stage of the voltage variation curve. Equation (6) shows that when the difference of the three sampled voltages is close to zero, the calculated value of ΔE_1 is nearly infinite because of signal interference in the actual application unlike in the ideal situation. This condition seriously affects the measurement result. A positive constant is set as C_a to determine if GC is too small by using the following rule.

$$\begin{cases} v_{11\_2} - v_{11\_3} < C_a \\ v_{22\_2} - v_{22\_3} < C_a \\ v_{12\_2} - v_{12\_3} > -C_a \\ v_{21\_2} - v_{21\_3} > -C_a \end{cases} \tag{13}$$

If the three sampled voltages satisfy Equation (13), then GC is too small to be non-negligible, and the traditional sampling method is adopted. If the three sampled voltages do not satisfy Equation (13), then GC is non-negligible, and the proposed three-point climbing algorithm is adopted.

4.3. Filter of E_1 and E_2

Δt is an invariant constant; τ_1 and τ_2 vary with GC, so E_1 and E_2 are also variations. Counter k is increased once every measurement period T_C, as shown in Figure 5. To make kth measurements E_1 and E_2 close to the actual value, the average value of M_1 and M_2 phases are taken according to Equation (6). $E_1(k)$ and $E_2(k)$ can be rewritten as

$$\begin{cases} E_1(k) = \left(\frac{v_{11\_1}(k) - v_{11\_2}(k)}{v_{11\_2}(k) - v_{11\_3}(k)} + \frac{v_{12\_1}(k) - v_{12\_2}(k)}{v_{12\_2}(k) - v_{12\_3}(k)}\right)/2 \\ E_2(k) = \left(\frac{v_{21\_1}(k) - v_{21\_2}(k)}{v_{21\_2}(k) - v_{21\_3}(k)} + \frac{v_{22\_1}(k) - v_{22\_2}(k)}{v_{22\_2}(k) - v_{22\_3}(k)}\right)/2 \end{cases} \tag{14}$$

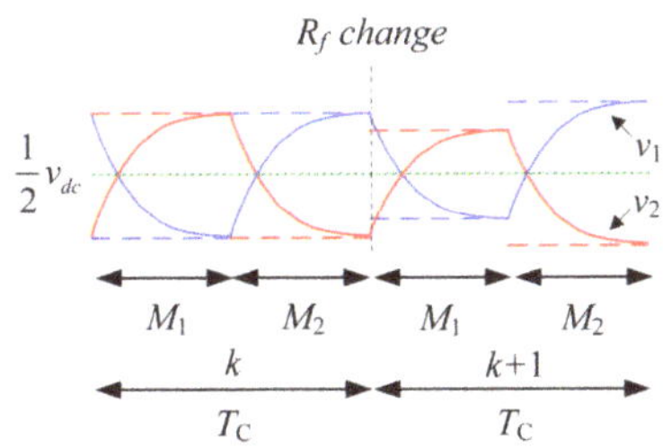

Figure 5. Waveform of the bridge voltage.

The estimated value of the kth E_1 and E_2 is set as $\hat{E}_1(k)$ and $\hat{E}_2(k)$, respectively, which can be obtained by the following first-order filter, where A is a filter coefficient that satisfies $0 < A < 1$.

$$\begin{cases} \hat{E}_1(k) = A\hat{E}_1(k-1) + (1-A)E_1(k) \\ \hat{E}_2(k) = A\hat{E}_2(k-1) + (1-A)E_2(k) \end{cases} \tag{15}$$

4.4. Correction of Sampled Value

The actual sampled voltage value shows a certain deviation from the expected value. The red sampled point in Figure 6 shows that the exponential function curve cannot be formed. The result calculated by the sampled value must be a large error. Therefore, to obtain satisfactory results, the sampled values must be corrected with a loop iterative correction method. $\hat{v}_{11\_1}(i)$,$\hat{v}_{11\_2}(i)$,$\hat{v}_{11\_3}(i)$,$\hat{v}_{12\_1}(i)$,$\hat{v}_{12\_2}(i)$,$\hat{v}_{12\_3}(i)$,$\hat{v}_{21\_1}(i)$,$\hat{v}_{21\_2}(i)$,$\hat{v}_{21\_3}(i)$,$\hat{v}_{22\_1}(i)$,$\hat{v}_{22\_2}(i)$, and $\widetilde{v}_{22\_3}(i)$ are set as the ith correction voltage values. After the 12 voltages of v_{11}, v_{12}, v_{21}, and v_{22} are sampled completely, the counter is set as $i = 0$. The 12 voltage values are substituted into the following equation, and the sampled value is used as the initial correction value.

$$\begin{cases} \hat{v}_{11\_1}(0) = v_{11\_1} \\ \hat{v}_{11\_2}(0) = v_{11\_2} \\ \hat{v}_{11\_3}(0) = v_{11\_3} \end{cases}, \begin{cases} \hat{v}_{12\_1}(0) = v_{12\_1} \\ \hat{v}_{12\_2}(0) = v_{12\_2} \\ \hat{v}_{12\_3}(0) = v_{12\_3} \end{cases} \begin{cases} \hat{v}_{21\_1}(0) = v_{21\_1} \\ \hat{v}_{21\_2}(0) = v_{21\_2} \\ \hat{v}_{21\_3}(0) = v_{21\_3} \end{cases}, \begin{cases} \hat{v}_{22\_1}(0) = v_{22\_1} \\ \hat{v}_{22\_2}(0) = v_{22\_2} \\ \hat{v}_{22\_3}(0) = v_{22\_3} \end{cases} \tag{16}$$

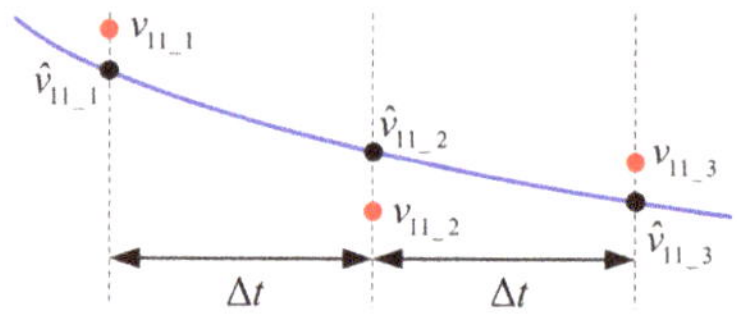

Figure 6. Actual sampling point.

The three sampled voltages of each group cannot form an exponential curve of $\hat{E}_1(k)$ due to the measurement error and ripple. To form the desired exponential curve, each voltage value can be estimated by the two other voltage values. $\tilde{v}_{11\_1}(i)$,$\tilde{v}_{11\_2}(i)$,$\tilde{v}_{11\_3}(i)$,$\tilde{v}_{12\_1}(i)$,$\tilde{v}_{12\_2}(i)$,$\tilde{v}_{12\_3}(i)$,$\tilde{v}_{21\_1}(i)$,$\tilde{v}_{21\_2}(i)$,$\tilde{v}_{21\_3}(i)$,$\tilde{v}_{22\_1}(i)$,$\tilde{v}_{22\_2}(i)$, and $\tilde{v}_{22\_3}(i)$ are set as the estimated voltage values. The estimated method can be derived from the following equation according to Equation (6).

$$\begin{cases} \tilde{v}_{11\_1}(i) = [\hat{v}_{11\_2}(i) - \hat{v}_{11\_3}(i)]\hat{E}_1(k)^{-1} + \hat{v}_{11\_2}(i) \\ \tilde{v}_{11\_2}(i) = [\hat{v}_{11\_1}(i)\hat{E}_1(k) + \hat{v}_{11\_3}(i)]/(1+\hat{E}_1(k)) \\ \tilde{v}_{11\_3}(i) = \hat{v}_{11\_2}(i) - [\hat{v}_{11\_1}(i) - \hat{v}_{11\_2}(i)]\hat{E}_1(k) \end{cases}, \begin{cases} \tilde{v}_{12\_1}(i) = [\hat{v}_{12\_2}(i) - \hat{v}_{12\_3}(i)]\hat{E}_1(k)^{-1} + \hat{v}_{12\_2}(i) \\ \tilde{v}_{12\_2}(i) = [\hat{v}_{12\_1}(i)\hat{E}_1(k) + \hat{v}_{12\_3}(i)]/(1+\hat{E}_1(k)) \\ \tilde{v}_{12\_3}(i) = \hat{v}_{12\_2}(i) - [\hat{v}_{12\_1}(i) - \hat{v}_{12\_2}(i)]\hat{E}_1(k) \end{cases}$$
$$\begin{cases} \tilde{v}_{21\_1}(i) = [\hat{v}_{21\_2}(i) - \hat{v}_{21\_3}(i)]\hat{E}_2(k)^{-1} + \hat{v}_{21\_2}(i) \\ \tilde{v}_{21\_2}(i) = [\hat{v}_{21\_1}(i)\hat{E}_2(k) + \hat{v}_{21\_3}(i)]/(1+\hat{E}_2(k)) \\ \tilde{v}_{21\_3}(i) = \hat{v}_{21\_2}(i) - [\hat{v}_{21\_1}(i) - \hat{v}_{21\_2}(i)]\hat{E}_2(k) \end{cases}, \begin{cases} \tilde{v}_{22\_1}(i) = [\hat{v}_{22\_2}(i) - \hat{v}_{22\_3}(i)]\hat{E}_2(k)^{-1} + \hat{v}_{22\_2}(i) \\ \tilde{v}_{22\_2}(i) = [\hat{v}_{22\_1}(i)\hat{E}_2(k) + \hat{v}_{22\_3}(i)]/(1+\hat{E}_2(k)) \\ \tilde{v}_{22\_3}(i) = \hat{v}_{22\_2}(i) - [\hat{v}_{22\_1}(i) - \hat{v}_{22\_2}(i)]\hat{E}_2(k) \end{cases} \quad (17)$$

The estimated value comparison rule is shown as

$$\begin{cases} \left|\hat{v}_{11\_1}(i) - \tilde{v}_{11\_1}(i)\right| < D, \left|\hat{v}_{11\_2}(i) - \tilde{v}_{11\_2}(i)\right| < D, \left|\hat{v}_{11\_3}(i) - \tilde{v}_{11\_3}(i)\right| < D \\ \left|\hat{v}_{12\_1}(i) - \tilde{v}_{12\_1}(i)\right| < D, \left|\hat{v}_{12\_2}(i) - \tilde{v}_{12\_2}(i)\right| < D, \left|\hat{v}_{12\_3}(i) - \tilde{v}_{12\_3}(i)\right| < D \end{cases}$$
$$\begin{cases} \left|\hat{v}_{21\_1}(i) - \tilde{v}_{21\_1}(i)\right| < D, \left|\hat{v}_{21\_2}(i) - \tilde{v}_{21\_2}(i)\right| < D, \left|\hat{v}_{21\_3}(i) - \tilde{v}_{21\_3}(i)\right| < D \\ \left|\hat{v}_{22\_1}(i) - \tilde{v}_{22\_1}(i)\right| < D, \left|\hat{v}_{22\_2}(i) - \tilde{v}_{22\_2}(i)\right| < D, \left|\hat{v}_{22\_3}(i) - \tilde{v}_{22\_3}(i)\right| < D \end{cases} \quad (18)$$

When Equation (18) is satisfied, the difference between the estimated value and the correction value is small, and the *i*th correction value is applied as the final correction value. Otherwise, the counter *i* is increased by 1, and further correction is be conducted as follows:

$$\begin{cases} \hat{v}_{11\_1}(i+1) = \tilde{v}_{11\_1}(i) + B[\hat{v}_{11\_1}(i) - \tilde{v}_{11\_1}(i)] \\ \hat{v}_{11\_2}(i+1) = \tilde{v}_{11\_2}(i) + B[\hat{v}_{11\_2}(i) - \tilde{v}_{11\_2}(i)] \\ \hat{v}_{11\_3}(i+1) = \tilde{v}_{11\_3}(i) + B[\hat{v}_{11\_3}(i) - \tilde{v}_{11\_3}(i)] \end{cases}, \begin{cases} \hat{v}_{12\_1}(i+1) = \tilde{v}_{12\_1}(i) + B[\hat{v}_{12\_1}(i) - \tilde{v}_{12\_1}(i)] \\ \hat{v}_{12\_2}(i+1) = \tilde{v}_{12\_2}(i) + B[\hat{v}_{12\_2}(i) - \tilde{v}_{12\_2}(i)] \\ \hat{v}_{12\_3}(i+1) = \tilde{v}_{12\_3}(i) + B[\hat{v}_{12\_3}(i) - \tilde{v}_{12\_3}(i)] \end{cases}$$
$$\begin{cases} \hat{v}_{21\_1}(i+1) = \tilde{v}_{21\_1}(i) + B[\hat{v}_{21\_1}(i) - \tilde{v}_{21\_1}(i)] \\ \hat{v}_{21\_2}(i+1) = \tilde{v}_{21\_2}(i) + B[\hat{v}_{21\_2}(i) - \tilde{v}_{21\_2}(i)] \\ \hat{v}_{21\_3}(i+1) = \tilde{v}_{21\_3}(i) + B[\hat{v}_{21\_3}(i) - \tilde{v}_{21\_3}(i)] \end{cases}, \begin{cases} \hat{v}_{22\_1}(i+1) = \tilde{v}_{22\_1}(i) + B[\hat{v}_{22\_1}(i) - \tilde{v}_{22\_1}(i)] \\ \hat{v}_{22\_2}(i+1) = \tilde{v}_{22\_2}(i) + B[\hat{v}_{22\_2}(i) - \tilde{v}_{22\_2}(i)] \\ \hat{v}_{22\_3}(i+1) = \tilde{v}_{22\_3}(i) + B[\hat{v}_{22\_3}(i) - \tilde{v}_{22\_3}(i)] \end{cases} \quad (19)$$

where B is the correction factor that satisfies $0 < B < 1$. Then, the results of Equation (19) are substituted into Equation (17). This method cycles back and forth until the difference between the estimated and correction values satisfies Equation (18). The cycle is then stopped, and the final correction value is outputted.

The overall software flow chart of the method is shown in Figure 7.

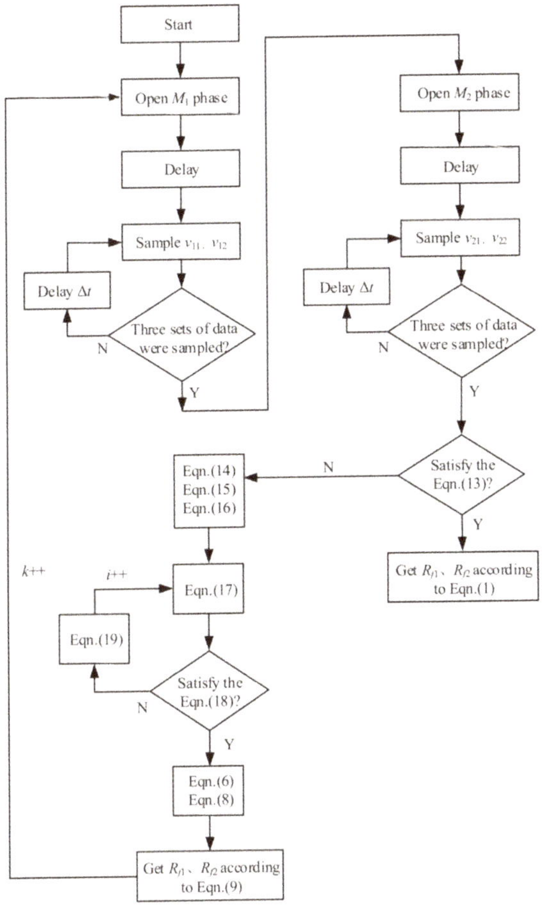

Figure 7. Software flow chart of the proposed method.

5. Comparative Study of Experimental Data

The DC-IM device with the unbalanced electric bridge method is created, and the schematic overview of application is shown in Figure 8. The MCU performs the switch Q_1 and Q_2, sample the voltage values of v_1 and v_2 stored in memory, then calculate the R_{f1} and R_{f2}, and output the result to the computer. The experiment table is shown in Figure 9. It includes the display interface, DC-IM device, voltage regulating device, insulation resistance selection switch, and GC selection switch. The DC-IM controller uses a PIC18F4580 single-chip microcomputer. The monitoring period is 0.2 s; that is, the switch action occurs every 0.1 s, so the sampling time should satisfy $(t_1 + 2\Delta t) < 0.1$ s. The positive grounding resistance is R_{f1} = 1000 kΩ, the negative grounding resistance is R_{f2} = 300 kΩ, the first sampling time t_1 = 0.01 s, the bridge resistance is R_a = 1000 kΩ, and R_b = 200 kΩ. The GC value is $C_Y = C_1 = C_2$. The grounding current waveform corresponding to different GCs is shown in Figure 10. The waveform of C_Y = 0.1 µF can be stabilized in a half cycle. The larger the value of C_Y is, the closer the waveform is to the triangle wave. Therefore, the traditional unbalanced bridge sampling method is used when C_Y < 0.1 µF, and the three-point climbing algorithm is used when C_Y > 0.1 µF.

The calculation results are compared by changing the different parameters, and relative error (RE%) is determined as

$$\text{RE\%} = |\text{ measured value} - \text{actual value }|/\text{measured value}. \tag{20}$$

Different parameters are applied in the proposed method. (1) DC voltage v_{dc} = 800 V, GC value C_Y = 0.1 μF, and the sampling time interval Δt is changed; the results are shown in Figure 11. The larger the sampling time interval is, the higher accuracy is. (2) DC voltage v_{dc} = 800 V, sampling time interval Δt = 0.04 s, and the value of C_Y is changed; the results are shown in Figure 12. The smaller C_Y is, the higher accuracy is. (3) Sampling interval Δt = 0.04 s, C_Y = 0.1 μF, and DC voltage v_{dc} is changed; the results are shown in Figure 13. The larger the DC voltage is, the higher accuracy is. When the DC voltage drops to below 200 V, the measurement accuracy is greatly reduced. (4) Under the premise that the parallel value of bridge resistance $R_a \parallel R_b$ is constant and the difference between R_a and R_b is changed; the results are shown in Table 2. The larger the difference between R_a and R_b is, the higher accuracy is. Time interval Δt increases, C_Y decreases, DC voltage v_{dc} increases, and the difference between R_a and R_b increases. These factors make the sampled voltage difference larger, which will reduce the error of the final results.

The proposed method is compared with the traditional method to verify the availability and superiority of the former. The monitoring time and relative error of the two methods are shown in Tables 3–5. The relative error is the larger one between R_{f1} and R_{f2}. Table 3 is the data at v_{dc} = 800 V, R_{f1} = 1000 kΩ, R_{f2} = 300 kΩ; Table 4 is the data at v_{dc} = 400 V, R_{f1} = 1000 kΩ, R_{f2} = 300 kΩ; Table 5 is the data at v_{dc} = 800 V, R_{f1} = 100 kΩ, R_{f2} = 100 kΩ. The traditional method needs to increase the monitoring time with a large value of GC because it should have a stable sample value after charging the GC. The proposed method has a fixed monitoring time due to the fixed three-point sampling. When GC is large and the proposed method is applied, the error is similar to using the traditional method. When GC is small, such as 10 nF in the table, and the traditional unbalanced bridge calculation method is applied automatically, the calculation results of two methods are almost the same. Overall, the experimental results are consistent with the theoretical conclusion.

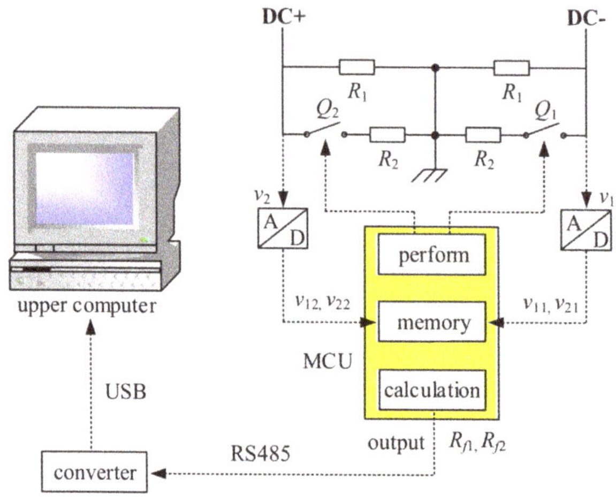

Figure 8. The schematic overview of application.

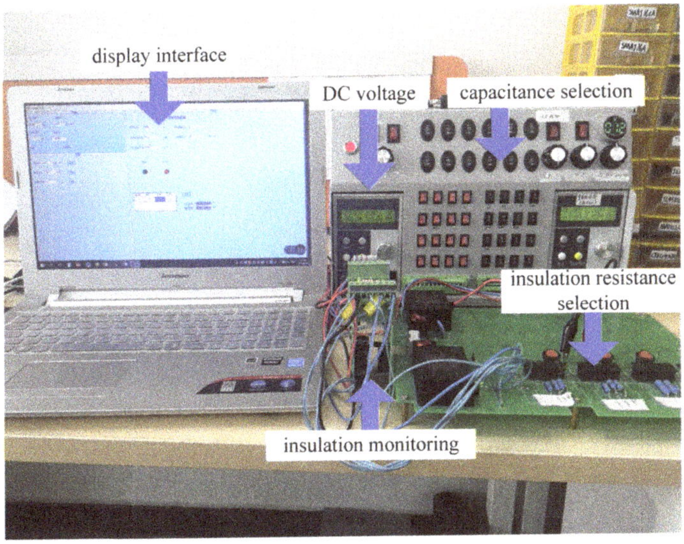

Figure 9. Display of the experiment table.

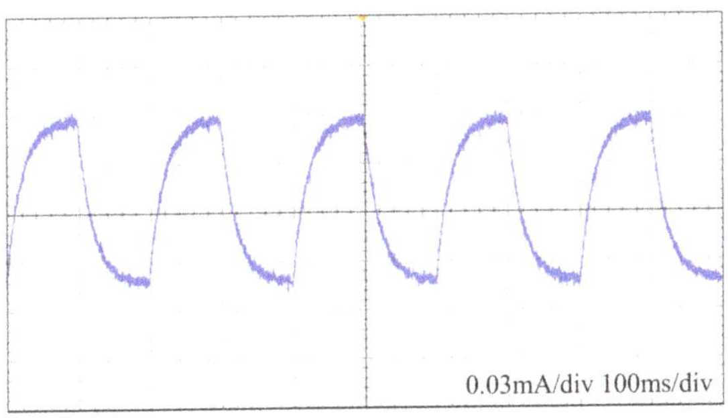

(**a**) $C_Y = 0.1$ μF

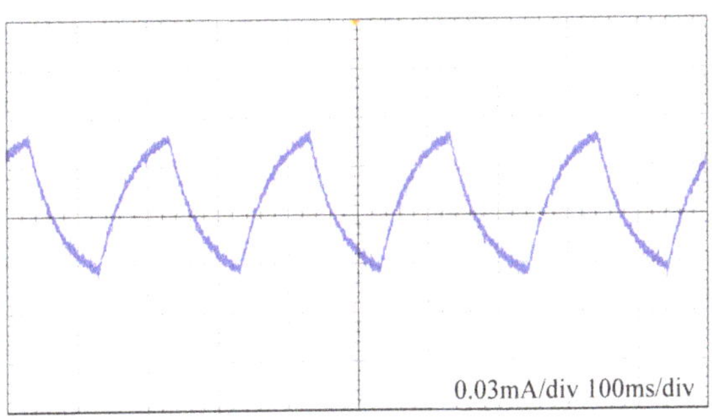

(**b**) $C_Y = 0.2$ μF

Figure 10. *Cont.*

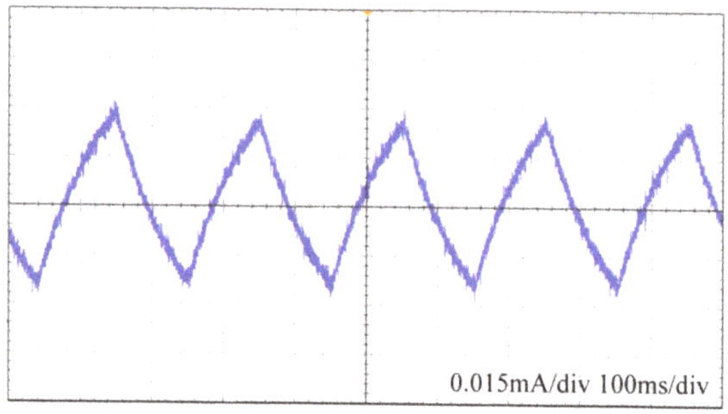

(c) $C_Y = 0.4$ μF

Figure 10. Waveform of grounding current with different values of C_Y. (**a**) $C_Y = 0.1$ μF. (**b**) $C_Y = 0.2$ μ. (**c**) $C_Y = 0.4$ μF.

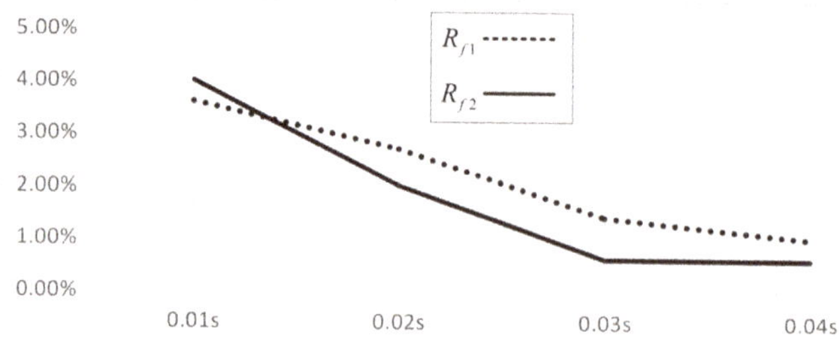

Figure 11. RE% with different sampling intervals.

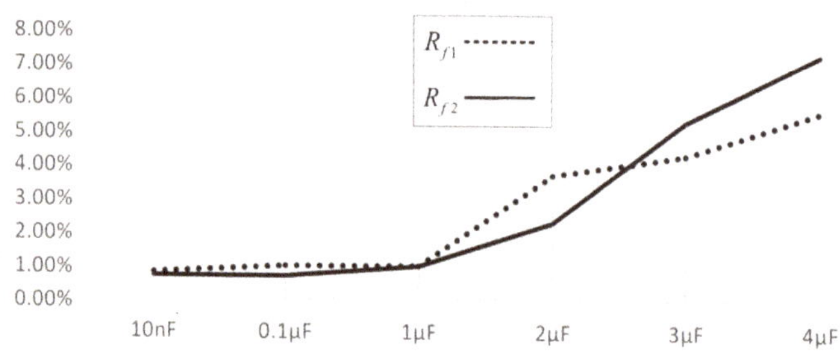

Figure 12. RE% with different values of CY.

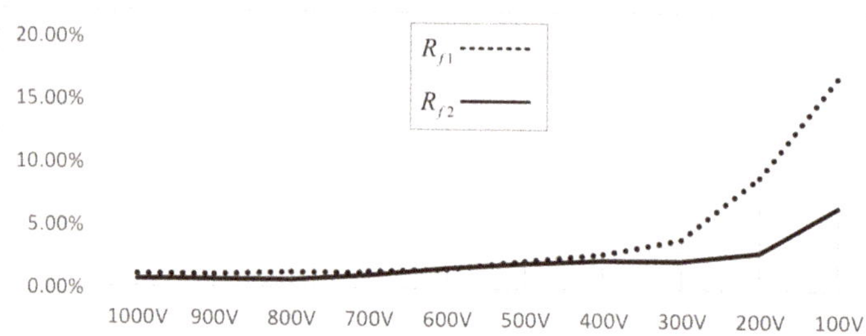

Figure 13. RE% with different DC voltages.

Table 2. Data results with different bridge resistors.

R_a (kΩ)	R_b (kΩ)	R_{f1} (kΩ)	RE% of R_{f1}	R_{f2} (kΩ)	RE% of R_{f2}
1000	200	1010	1.0%	302	0.6%
800	250	1014	1.4%	305	1.6%
600	300	1020	2.0%	309	3.0%
500	350	1045	4.5%	316	5.3%

Table 3. Comparison of monitoring data in v_{dc} = 800 V, R_{f1} = 1000 kΩ, R_{f2} = 300 kΩ.

	Traditional Method		Proposed Method	
C_Y	Monitoring Time	RE%	Monitoring Time	RE%
10 nF	0.2s	0.8%	0.2s	0.8%
0.1 μF	0.3s	1.0%	0.2s	1.2%
1 μF	1.35s	2.4%	0.2s	1.8%
2 μF	2.51s	3.5%	0.2s	3.7%
3 μF	3.67s	5%	0.2s	5.3%
4 μF	4.83s	6.6%	0.2s	7.3%

Table 4. Comparison of monitoring data in v_{dc} = 400 V, R_{f1} = 1000 kΩ, R_{f2} = 300 kΩ.

	Traditional Method		Proposed Method	
C_Y	Monitoring Time	RE%	Monitoring Time	RE%
10 nF	0.2s	1.3%	0.2s	1.3%
0.1 μF	0.29s	1.7%	0.2s	2.8%
1 μF	0.34s	3.2%	0.2s	4.1%
2 μF	2.49s	5.8%	0.2s	6.2%
3 μF	3.64s	8.5%	0.2s	9%
4 μF	4.79s	11%	0.2s	12%

Table 5. Comparison of monitoring data in v_{dc} = 800 V, R_{f1} = 100 kΩ, R_{f2} = 100 kΩ.

	Traditional Method		Proposed Method	
C_Y	Monitoring Time	RE%	Monitoring Time	RE%
10 nF	0.2s	0.7%	0.2s	0.7%
0.1 μF	0.23s	1.0%	0.2s	0.9%
1 μF	0.65s	2.2%	0.2s	1.6%
2 μF	1.11s	3.4%	0.2s	2.4%
3 μF	1.57s	4.7%	0.2s	4.1%
4 μF	2s	6.3%	0.2s	5.2%

6. Conclusions

For the DC-IM circuit of EVs, the traditional unbalanced electric bridge method switches the positive and negative bridge resistances and calculates the insulation resistance value by sampling the positive and negative bridge voltages. However, when the DC positive and negative poles have GC, the bridge voltages must be sampled after the capacitor is charged completely; thus, the measurement time is very long. This study proposes a novel method of DC-IM using a three-point climbing algorithm. The insulation resistance can be calculated by sampling the voltage of the positive and negative bridges three times and keeping the sampling interval equal. Moreover, the method filters and automatically corrects the three sampling voltages, which can improve the accuracy of the calculation results. Combined with experimental data, the following conclusions can be drawn: (1) The advantage of proposed method can perform a faster time and maintain a constant monitoring period compared with the traditional method. (2) The restriction of proposed method only apply in larger GC situation. If GC is small, the traditional method could be used. (3) The characteristics of proposed method:

Increasing the sampling interval, increasing the difference between R_a and R_b, increasing DC voltage v_{dc}, all make the results more accurate. Overall, the proposed method can be applied to some practical industrial applications. The future work is to study how to set the constant C_a to determine which method to use or find a different rule to generally judge the value of GC, and study a method to reduce error when v_{dc} is constant changing.

Author Contributions: J.D. and T.Q.Z. provided the method and solution; Y.Y. and H.Z. performed the experiments; J.D. and Y.Y. analyzed the data; J.D. and Y.Z. wrote the paper; T.Q.Z. and H.L. checked paper.

Funding: This research received no external funding.

Conflicts of Interest: The authors declare no conflict of interest.

References

1. Hou, Y.P.; Zhou, W.; Shen, C.Y. Experimental investigation of gas-tightness and electrical insulation of fuel cell stack under strengthened road vibrating conditions. *Int. J. Hydrogen Energy* **2011**, *36*, 13763–13768. [CrossRef]
2. Gyftakis, K.N.; Sumislawska, M.; Kavanagh, D.F.; Howey, D.A.; McCulloch, M.D. Dielectric Characteristics of Electric Vehicle Traction Motor Winding Insulation Under Thermal Aging. *IEEE Trans. Ind. Appl.* **2016**, *52*, 1398–1404.
3. Dannier, A. Overview of Main Electric Subsystems of Zero-Emission Vehicles. *Propuls. Syst.* **2019**, 1–31. [CrossRef]
4. Vahedipour-Dahraie, M.; Rashidizaheh-Kermani, H.; Najafi, H.R.; Anvari-Moghaddam, A.; Guerrero, J.M. Coordination of EVs Participation for Load Frequency Control in Isolated Microgrids. *Appl. Sci.* **2017**, *7*, 539. [CrossRef]
5. Rafique, M.K.; Khan, S.U.; Zaman, M.S.U.; Mehmood, K.K.; Haider, Z.M.; Ali Bukhar, S.B.; Kim, C. An Intelligent Hybrid Energy Management System for a Smart House Considering Bidirectional Power Flow and Various EV Charging Techniques. *Appl. Sci.* **2019**, *9*, 1658. [CrossRef]
6. Kim, J.; Kim, H.; Cho, Y.; Kim, H.; Cho, J. Application of a DC Distribution System in Korea: A Case Study of the LVDC Project. *Appl. Sci.* **2019**, *9*, 1074. [CrossRef]
7. Holger Potdevin, D.-I. Insulation Monitoring in High Voltage Systems for Hybrid and Electric Vehicles. High Voltage Safety. *ATZ Elektron.* **2009**, *4*, 28–31. [CrossRef]
8. Pan, X.F.; Rinkleff, T.; Willmann, B.; Vick, R. PSpice simulation of surge testing for electrical vehicles. In Proceedings of the 2012 International Symposium on Electromagnetic Compatibility (EMC EUROPE), Rome, Italy, 17–21 September 2012.
9. Liu, Y.-C.; Lin, C.-Y. Insulation fault detection circuit for ungrounded DC power supply systems. In Proceedings of the 2012 IEEE Sensors, Taipei, Taiwan, 28–31 October 2012.
10. Zhao, C.Y.; Jia, X.F.; Hao, Z.F. The new method of monitoring DC system insulation on-line. In Proceedings of the 27th Annual Conference of the IEEE Industrial Electronics Society, IECON'01, Denver, CO, USA, 29 November–2 December 2001.
11. Jiang, J.S.; Ji, H. Study of Insulation Monitoring Device for DC System Based on Multi-switch Combination. In Proceedings of the Second International Symposium on Computational Intelligence and Design, ISCID'09, Changsha, China, 12–14 December 2009.
12. Li, J.X.; Wu, Z.Q.; Fan, Y.Q.; Wang, Y.Y.; Jiang, J.C. Research on Insulation Resistance On-Line Monitoring for Electric Vehicle. In Proceedings of the 2005 International Conference on Electrical Machines and Systems, Nanjing, China, 27–29 September 2005.
13. Wei, X.Y.; Bi, L.; Sun, Z.C. A method of insulation failure detection on electric vehicle based on FPGA. In Proceedings of the 2008 IEEE Vehicle Power and Propulsion Conference, VPPC'08, Harbin, China, 3–5 September 2008.
14. Zheng, M.X. The Design of Insulation Monitoring Device for Electric Vehicle's Battery Pack. *Adv. Mater. Res.* **2013**, *718–720*, 1422–1428. [CrossRef]
15. Bao, Y.; Jiang, J.C.; Zhang, W.G.; Wang, J.Y.; Wen, J.P. Novel On-line Insulation Supervising Method for DC System. *High Volt. Eng.* **2011**, *37*, 333–337.

16. Wu, Z.J.; Wang, L.F. A novel insulation resistance monitoring device for Hybrid Electric Vehicle. In Proceedings of the 2008 IEEE Vehicle Power and Propulsion Conference, VPPC'08, Harbin, China, 3–5 September 2008.
17. Neti, P.; Grubic, S. Online broadband insulation spectroscopy of induction machines using signal injection. *IEEE Trans. Ind. Appl.* **2017**, *53*, 1054–1062. [CrossRef]
18. Chiang, Y.-H.; Sean, W.-Y.; Huang, C.-Y.; Chiang Hsieh, L.-H. Adaptive control for estimating insulation resistance of high-voltage battery system in electric vehicles. *Environ. Prog. Sustain. Energy* **2017**, *36*, 1882–1887. [CrossRef]
19. Li, D.H.; Jia, W. Application of Chaos Theory in Grounaing Fault Detection of DC Power Supply System. *High Volt. Eng.* **2005**, *31*, 79–82.
20. Li, D.H.; Shi, L.T. An Overall Scheme for Ground Fault Detection in DC Systems of Power Plants and Substations. *Power Syst. Technol.* **2005**, *29*, 56–59.
21. Song, C.X.; Shao, Y.L.; Song, S.X.; Peng, S.L.; Zhou, F.; Chang, C.; Wang, D. Insulation Resistance Monitoring Algorithm for Battery Pack in Electric Vehicle Based on Extended Kalman Filtering. *Energies* **2017**, *10*, 714. [CrossRef]

applied sciences

MDPI

Article

Improved Control of Forest Microgrids with Hybrid Complementary Energy Storage

Ming Yu, Junguo Zhang * and Hanxing Liu

School of Technology, Beijing Forestry University, Beijing 100083, China; yuming@bjfu.edu.cn (M.Y.); liuhanxing@bjfu.edu.cn (H.L.)
* Correspondence: zhangjunguo@bjfu.edu.cn; Tel.: +86-10-6233-6398

Received: 6 June 2019; Accepted: 19 June 2019; Published: 20 June 2019

check for updates

Featured Application: The improved control method using complementary energy storage units can enhance the transient stability and operation reliability of microgrids in forests, providing theoretical support for the construction of forest microgrids.

Abstract: In order to improve the power quality and the fault ride-through capability of islanded forest microgrids, a hybrid complementary energy storage control method is proposed. In this method, mode-based sectional coordinated control is adopted as the basic control scheme, whereas control of the hybrid energy storage, which includes the battery, the supercapacitor, and the wind turbine, utilizes the improved strategy. According to the characteristics of the energy storage units, adaptive control of batteries and supercapacitors are adopted to smooth the low-frequency power fluctuation in the long-term and to suppress the high-frequency component separately, in which predictive control of the converters is utilized to achieve rapid regulation. Furthermore, as a third energy storage unit, the wind power unit was investigated, utilizing the large rotating kinetic energy of the wind turbine to temporally suppress huge power disturbance and avoid load shedding. To verify the effectiveness of the proposed coordination control with hybrid complementary energy storage, simulations of the islanded DC microgrid in forest area were conducted in MATLAB/Simulink, with the results showing that, by utilizing the improved control method, the transient operation characteristics of the system were effectively enhanced.

Keywords: forest microgrid; biomass energy; operation mode-based sectional coordinated control; hybrid energy storage; predictive control; inertia improvement

1. Introduction

With the increasing efforts of forestry encouragement and protection policy, the forest area and corresponding stock volume in China have achieved double growth. The carbon sink capacity of the forest has been greatly enhanced, and its ecological barrier function is stable, with good momentum for growth [1]. However, the gradual expansion of forest area has presented a major challenge to the power supply and distribution systems in forest areas. Capacity expansion of forest power networks always covers large areas with long transmission lines, which leads to huge construction investment and high loss in the power line. In addition, the maintenance area of the forestry power grid is expanded, and the corresponding human, material, and financial resources invested in operation and maintenance are increased. Therefore, the management pressure of power companies has risen remarkably [2–4].

As a low-cost utilization method of renewable energy, the microgrid, which is based on renewable energy, has the advantages of small initial investment, reliable power supply, small transmission loss, and flexible operation. It is a useful supplement to the utility grid, and helps to promote sustainable development of energy [5]. The microgrid can not only operate in grid-connected mode, but also in islanded mode, which makes it suitable for niche applications such as power supplies in isolated forest locations [6]. By effectively utilizing the abundant wind energy, biomass energy, and other natural energy locally, clean and efficient power can be provided for the loads in forests, such as monitoring and alarm systems, wood processing machinery, or electrical equipment on watchtowers. Therefore, the problems of power construction, operation, and management in remote forest areas can be solved.

Being different from islanded microgrids constructed in environments such as islands, microgrids in forests have their own ecological particularity. A large amount of residue from forest clearing, logging, and wood processing cannot usually be cleaned up and transported in time, which is a hidden hazard that may trigger forest fires [7]. Scientists in the United Kingdom and Sweden have proposed using the surplus produced by cleaning and afforesting to generate economic benefits, so as to promote the restoration of the forest ecosystem. American scientists have also paid great attention to wood power generation, and invested plenty of money on related research [8,9]. Using forestry residues for power generation can not only dispose of forest waste nearby, saving transportation costs and avoiding fires, but can also provide natural and environmentally friendly energy support for power loads in remote forest areas.

The forest microgrid in this paper, which is composed of a biomass power generation unit, wind power unit, and energy storage units, adopts a DC bus to improve stability and reliability during operation. In addition, in a DC microgrid there are no such problems as frequency deviation, reactive current circulation, and power angle stability, which are normally encountered in AC systems. The distributed generators and energy storage units in islanded DC microgrids are connected to the DC bus through power electronic converters, with asynchronous motors providing rotational kinetic energy as a natural inertial support. Therefore, the forest microgrid is an intrinsic small inertial system. In the case of large power disturbance, a sudden change of DC voltage will pose a threat to the voltage-sensitive load, leading to load shedding. In addition, fast variation of DC voltage caused by random fluctuations of distributed energy will also affect the quality of the output power. Therefore, it is of practical significance to investigate advanced control schemes of DC microgrids, in order to improve the transient response and ensure its safe and stable operation.

Scholars have proposed various control schemes of energy storage devices to suppress the power fluctuations of microgrids [10–19]. In Reference [10], a synergistic operation between converters of battery energy storage and a photovoltaic generator to assist management of microgrids is presented and in Reference [11], current-controlled bidirectional DC/DC converters were applied to connect each lithium ion battery bank as well. Although the control methods of converters in the above literature can ensure the basic stability of corresponding systems, time delay is inevitable, in which traditional proportional integral adjustment or corresponding improved forms based on deviation of the controlled variables are utilized. An experimental investigation of an energy storage unit which incorporates electric energy storage in the form of hybrid capacitors and hydraulic energy storage in the form of pressure vessels in a photovoltaic powered seawater reverse osmosis desalination system was proposed by Karavas, whereas there was only innovation in the energy storage form [12]. In Reference [13], a wireless droop control method for distributed energy storage units in AC microgrids is presented, which employs the SoC-based droop control method locally to prolong the service life of the energy storage, and in Reference [14], the energy storage was scheduled to work in a grid supportive manner, with a grid adaptive power management strategy being formulated to generate current references for energy storage systems and microgrid-connected converters. Both methods emphasize different priorities on various control targets, with indifference to improving control speed to enhance transient stability, which is critical in islanded microgrids. In References [15–17], optimal controls for microgrids with hybrid energy storage system were carried out using model predictive control (MPC), which allowed maximization of the economic benefits of the microgrids, minimizing the degradation causes of storage systems, or fulfilling other, different system constraints. The focuses of these papers are economical schedules in the upper control level, having nothing to do with improvement of the response speed or the transient characteristics of microgrids. A hierarchical control of a hybrid energy storage system, composed of both centralized and distributed control, was proposed in [18], and a method consisting of a virtual resistance droop controller and a virtual capacitance droop controller for energy storages with complementary characteristics was proposed in [19]. Although more accurate current references for DC converters were generated, the control speed of converters did not improve, which weakened the effectiveness of the control strategy. In References [20,21], the authors used a two-level control scheme to control the charge/discharge power of the storage unit, in which the reference in the first level was obtained through a robust optimal power management system. Moreover, the research on inertial control of the energy storage system was limited to adding a supercapacitor into the microgrid to improve the equivalent inertia, whereas the effect of disturbance suppression was not obvious [22–24].

In this paper, a coordination control strategy of hybrid complementary energy storage in a forest microgrid is proposed, to improve its transient operation stability and the fault ride-through capability. The battery, the supercapacitor, and the wind turbine energy storage unit constitute the hybrid energy storage system, all of which undertake different tasks in maintaining power balance under various operation modes. To better ensure the control speed and efficiency of the battery and the supercapacitor under power fluctuation conditions, the adaptive droop coefficients and predictive converter control method are firstly proposed. In the following, an inertia enhancement control strategy of the wind turbine, utilizing its rotating kinetic energy, was investigated to improve its fault ride-through capacity in urgent conditions, such as load shedding. In addition, overall control of the hybrid complementary energy storage system was studied based on the fundamental DC voltage sectional control to coordinate various energy storage units.

This paper is organized as follows. Section 2 presents the basic operation mode and coordination control of DC microgrid. Section 3 describes the hybrid complementary energy storage system, while the overall control of the forest microgrid is discussed in Section 4. Simulations are presented in Section 5.

2. Operation Mode and Coordination Control of Microgrid in Forest Area

2.1. Structure of the Forest Microgrid

The forest microgrid adopts a DC bus due to the advantages of flexible control, low line loss, and high conversion efficiency, which determine it to be a more ideal connection pattern than its AC counterpart. Therefore, it is conducive to the flexible access of renewable energy and the realization of intelligent power supply in remote forest areas.

In this paper, the forest microgrid, which is composed of the biomass power unit, the wind power unit, and energy storage units, adopts multiple energy sources in complementary forms. As the microgrid is designed for remote forest areas, which are hard to access by the utility grid, it operates in islanded (off-grid) mode. Moreover, radiation topology is adopted by the system with simple structure and low construction cost, as is shown in Figure 1.

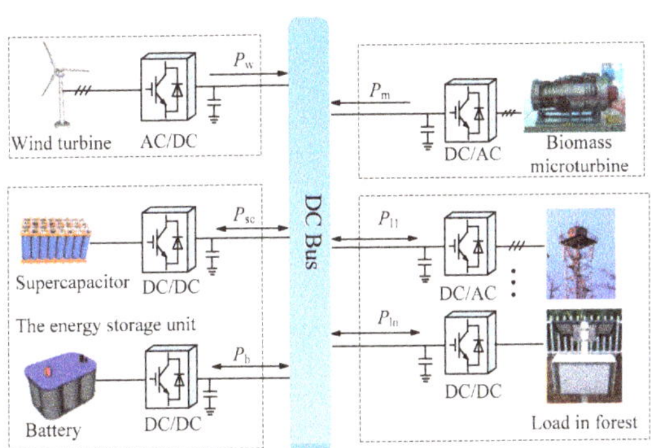

Figure 1. System structure of DC microgrid with wind turbine, energy storage, and biomass energy.

When the system operates normally, the wind power unit provides green energy to support the microgrid. The biomass power generation unit generally operates in hot standby mode, while the supercapacitor, which cooperates with the biomass unit or batteries, is utilized to eliminate fast power disturbance of the system, based on its characteristic of fast charging and discharging. In addition, the battery energy storage system is an emergency power supply unit, which can realize bidirectional power flow through the DC/DC converter. According to the operation characteristics of each unit, the operation mode of the forest DC microgrid can be divided into the following four parts.

2.2. Operation Mode of the Forest Microgrid

2.2.1. Free Mode

Being different from a grid-connected microgrid, the forest microgrid is isolated from the utility grid and is an energy self-sufficient system. In free mode, the system operates normally under small power disturbance, with DC voltage of the microgrid being in the interval of [0.98, 1.02] pu (per-unit value). The wind turbine unit carries out maximum power point tracking, while the biomass power generation unit takes in charge of suppressing the power disturbance by regulating its output power. Furthermore, the supercapacitor performs high frequency power smoothing and the battery is charged with a small constant current when the voltage is above its rated value.

2.2.2. Emergency Mode

When the wind is relatively weak or the output of the biomass power generation unit is limited, the DC voltage of the microgrid will drop below 0.98 pu. Under these circumstances, the energy storage system discharges to supplement the power gap. When the output of the wind power generation system is so high that the DC voltage rises above 1.02 pu, the biomass power generation unit withdraws automatically. In this case, if the system power is still excessive, the battery can be charged only when its state of charge is below the upper limit.

2.2.3. Power Limiting Mode

When the wind power is greater than the aggregate demand of the load and the energy storage systems, maintaining maximum power point tracking of wind power unit will lead to a sharp rise in DC bus voltage to above 1.05 pu. Therefore, inertia control of the wind turbine will be conducted, which temporarily stores the excess power in the form of kinetic energy of the wind turbine. If the working condition continues, power-limiting control of the wind power unit should be carried out to maintain stability of the system.

2.2.4. Load-Shedding Mode

When the power deficiency lasts and the load demand is still unable to be met, even if the energy storage is depleted, the wind turbine releases its kinetic energy to temporarily extend the operation time of the load. Under long working conditions of low power input, the load needs to be shed according to priority.

2.3. Coordination Control

With no frequency and reactive power in the DC microgrid, the DC voltage fluctuation is mainly caused by the uncertainty of system input power [25]. As the DC bus voltage reflects the internal power state of the microgrid and is a sign of system stability, power–voltage regulation is the core of stability control, as well as energy management. This paper utilizes DC bus voltage as the basis for operation mode, switching of each unit to effectively simplify system control.

In this method, the various units of the system are divided into a slack terminal, which is responsible for adjusting power, and power terminals, with constant output in different working modes, as is shown in Figure 2. They are not static, and switch with the change of the operating state of the system. The corresponding switching diagram of control mode is shown in Figure 3. For example, under free mode and emergency mode, the primary power regulating units of the system are the biomass power generation unit and the energy storage unit, respectively. In power limiting mode, the corresponding power control unit is the wind power system.

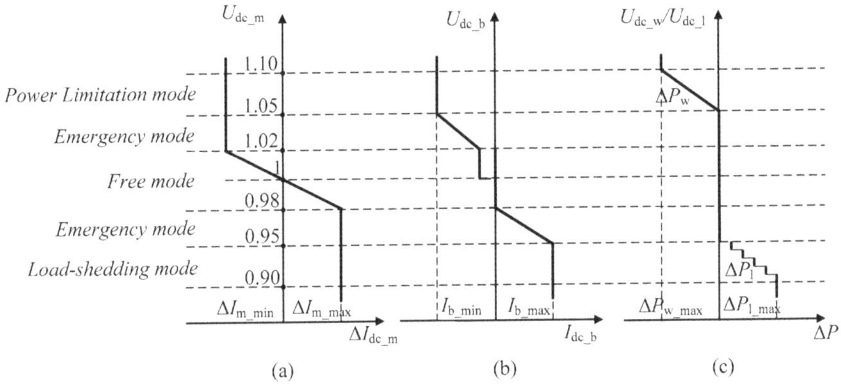

Figure 2. Basic principle of operation mode-based sectional coordinated control strategy for DC microgrid in forest: (**a**) Biomass power generation unit; (**b**) Energy storage unit; (**c**) Wind power unit/Load.

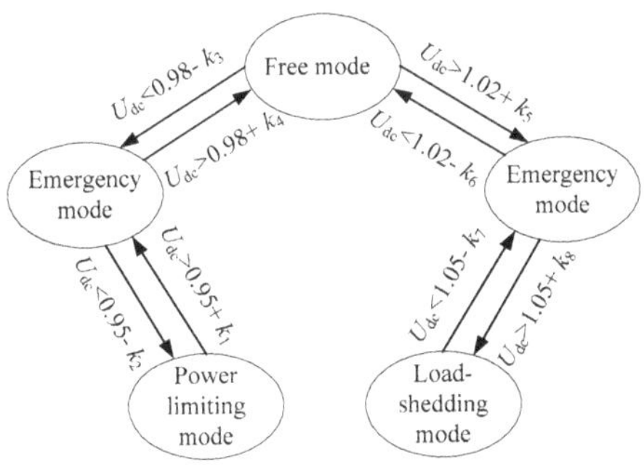

Figure 3. State switching diagram of system control mode. k_i (i = 1 ... 8): the hysteresis bandwidth in different modes.

In addition, the supercapacitor functions under the condition of rapid power fluctuation in free mode and emergency mode, while kinetic energy regulation of the wind turbine acts under large power disturbances in power limiting mode as well as load shedding mode.

In the DC microgrid, as the voltage value at the DC bus of each unit makes very little difference, the working mode of each unit can be determined only by detecting the DC voltage at the outlet of each converter. Consequently, there is no need to establish a complex communication system by utilizing an operation mode-based sectional coordinated control strategy. With the characteristic of plug and play, the coordinated control method has the advantage of simplicity and reliability. In addition, it boasts the feature of self-adaption besides coordination of different units. Due to the fact that the operation mode of the system is divided based on DC voltage, and under each mode power regulation is conducted by the most suitable controllable unit, the stability of the islanded system can be ensured.

3. Hybrid Complementary Energy Storage System

3.1. The Necessity of the Hybrid Complementary Energy Storage

The DC microgrid based on power electronic converters is a system with small inertia, which presents large fluctuation when being subjected to power disturbance. Although microgrids are normally equipped with an energy storage device to ensure stability of the system, when confronted with temporary high power fluctuations, the control system may fail to respond in time due to control delay. In this condition, load shedding or system protection will be caused. By coordinating the three kinds of energy storage units proposed in the DC forest area microgrid, adopting the appropriate energy storage unit to suppress system disturbances under different operation modes or disturbance conditions, the response characteristics of the system can be improved. Accordingly, as the transient stability of the system is enhanced, large fluctuations of DC voltage can be restrained.

As batteries have the advantage of high energy density, they are suitable for long term energy storage to compensate for power gaps when system input is limited in emergency mode. However, their charge–discharge efficiency is low, which is not recommended for frequent action to suppress power fluctuations. Contrarily, supercapacitors have the merits of high power density and long cycle life, as well as high charge and discharge efficiency [26]. Therefore, they are appropriate for suppressing the power fluctuations with high frequency and small amplitude that are typical in free mode and emergency mode, to assist the biomass unit in power regulation. In addition, by exploiting the large rotating kinetic energy of wind turbine, large disturbances of the system can be suppressed, which can effectively improve the fault ride-through capability of the microgrid and avoid load shedding.

3.2. Improvement in Traditional Control of Batteries and Supercapacitors

The reference current absorbed or released by the energy storage system is divided into a low frequency part and a high frequency part through the low pass filter (LPF). As batteries boast high energy density and large energy storage, they are used as long-term devices for power balance by absorbing or releasing low frequency power, while supercapacitors are utilized for suppressing high frequency disturbances.

The power disturbances that lead the system to emergency mode always give rise to large change rates of DC voltage ($\mathrm{d}U_{dc}/\mathrm{d}t$). Therefore, $\mathrm{d}U_{dc}/\mathrm{d}t$ can be utilized to adaptively modify the droop coefficient of the energy storage converter. When it exceeds the preset threshold C, the droop coefficient k_i varies adaptively to improve the response speed, as shown in Equation (1).

$$k_i = \begin{cases} k_{n\_i} - m_1\left(\left|\frac{\mathrm{d}U_{dc}}{\mathrm{d}t}\right|\right)^{m_2} & ,\text{for}\left|\frac{\mathrm{d}U_{dc}}{\mathrm{d}t}\right| \geq C \\ k_{n\_i} & ,\text{for}\left|\frac{\mathrm{d}U_{dc}}{\mathrm{d}t}\right| < C \end{cases}, \quad (1)$$

where k_{n_i} is the droop coefficient of the converter in the original control method. Constants m_1 and m_2 are based on the rated capacity of microgrid converter and the maximum allowable deviation of DC voltage. m_1 is determined by the following Equations.

$$m_1 = \frac{k_{n\_i} - k_{i\_min}}{\left(\left|\frac{\mathrm{d}U_{dc}}{\mathrm{d}t}\right|_{max}\right)^{m_2}}, \quad (2)$$

$$k_{i\_min} = \frac{\Delta U_{dci}}{\Delta I_{i\_max}}, \quad (3)$$

where k_{i_min} is the minimum droop gain, the selection of which is to prevent the output power of the converter from exceeding its maximum limit. ΔU_{dci} is the variation of DC voltage corresponding to the maximum current limit (ΔI_{i_max}) of the convertor, and is intended for setting the allowable voltage variation corresponding to a specific converter.

The value of m_1 depends on the output capacity of the converter represented by the maximum voltage change rate $|\mathrm{d}U_{dc}/\mathrm{d}t|_{max}$ and k_{i_min}. The power support capability of the energy storage converter increases in the same direction with m_1. A small m_1 may result in transient overshoot of DC voltage, while a large one may cause power oscillation of the system. Therefore, it is critical to select an appropriate m_1. Similarly, a smaller m_2 corresponds to smaller droop coefficient of the converter, which implies more power available at the instant of disturbance, providing reliable energy support for the system.

The generation procedure of the reference value of the energy storage converter with an adaptive droop coefficient is shown in Figure 4. When the system operates steadily, the change rate of DC voltage is less than the set threshold C_i. Therefore, the output of the comparator is 0, and the converter operates with the original droop coefficient. When the system is disturbed rapidly with large amplitudes, which leads the system to emergency mode, the change rate of DC voltage may exceed the set threshold. Under this circumstance, the output of the comparator is 1 and the dynamic droop coefficient of the converter is adopted, which smooths the DC voltage.

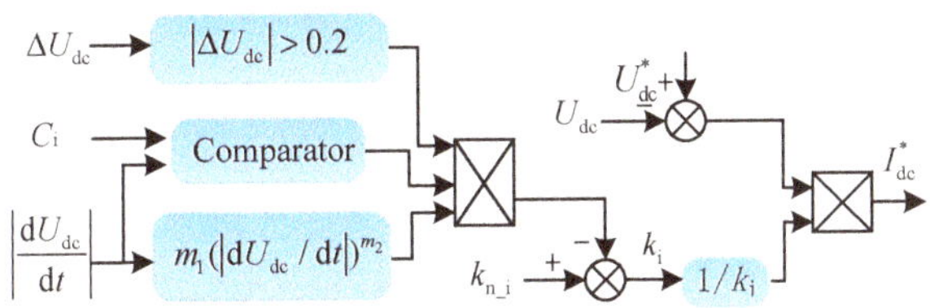

Figure 4. Principle diagram of self-adaptive inertia droop control.

3.3. Predictive Converter Control Method

Achieving fast smoothing of the system power requires fast converter control, whereas adopting the routine method based on real time value of DC/DC converters may weaken the effect of the self-adaptive droop control due to the lag regulation of current. Therefore, a predictive control method is proposed for the hybrid energy storage system to improve corresponding regulation speed. The converters of the supercapacitor and battery adopt the same Buck/Boost topology, as shown in Figure 5, with the switches V_1 and V_2 operating complementarily [27]. When the DC voltage in the microgrid is lower than the switching threshold, the energy storage unit discharges, in which condition the converter works in Boost mode. Contrarily, when there is power surplus and the energy storage unit needs to be charged, the converter switches to Buck mode.

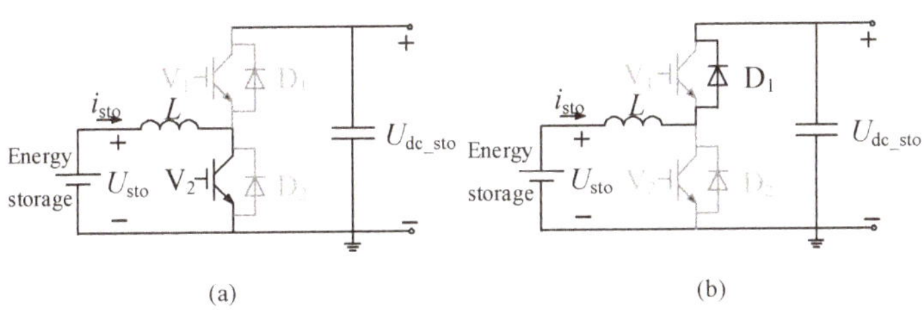

Figure 5. Topology and working mode of bi-directional DC/DC converter in Boost mode: (**a**) Conduction period of V_2; (**b**) Shutdown period of V_2.

The converter of the energy storage system usually adopts a double closed-loop control structure, with the outer-loop being voltage control and the inner loop current control. Being different from the commonly used PI control method, the predictive control method based on converter model collects the current state variables of the system and calculates the predictive current value at the next moment through the predictive model. The switching action of the converter is then chosen by minimizing the deviation between the predicted current value and the reference one [28]. This method adopts active predictive control instead of passive feedback regulation, which effectively avoids the time lag in traditional PI-based current regulation within the inner loop. Therefore, the predictive method is suitable for the occasion of voltage regulation, which requires high control speed.

3.3.1. Current Prediction Model of Converter

In Boost mode, V_2 and the anti-parallel freewheeling diode D_1 are in working state. During the turn-on stage of switch V_2 ($Sc_2 = 1$), the energy storage unit charges the inductor L through V_2, as shown in Figure 5a. Equation (4) can be obtained from its equivalent circuit, which is further discretized to obtain the predicted current value at the $k + 1$th moment, shown in Equation (5).

$$LdI_{sto}/dt = U_{sto}, \tag{4}$$

$$i_{sto}(k+1) = T_s u_{sto}(k)/L + i_{sto}(k), \tag{5}$$

where $u_{sto}(k)$ denotes the voltage detected at both ends of the energy storage unit (the battery or the supercapacitor) at the kth moment; $i_{sto}(k)$ represents the current of the energy storage unit, which flowing through inductance L at the kth moment and T_s is the sampling period.

During the switching off stage of V_2 ($S_{C2} = 0$), the electromagnetic energy stored in inductance L is released to the DC side through the anti-parallel freewheeling diode D_1. The following formula can be obtained from the circuit shown in Figure 5b.

$$LdI_{sto}/dt = U_{sto} - U_{dc\_sto}. \tag{6}$$

Equation (6) is discretized and the predicted current is as follows.

$$i_{sto}(k+1) = T_s[u_{sto}(k) - u_{dc\_sto}(k)]/L + i_{sto}(k), \tag{7}$$

where $u_{dc\_sto}(k)$ denotes the DC side voltage of the converter at the kth moment.

Similarly, when the bidirectional DC/DC converter operates in Buck mode, its prediction model of current is established as follows.

$$\begin{cases} i_{sto}(k+1) = T_s[-u_{sto}(k) + u_{dc\_sto}(k)]/L + i_{sto}(k), & (S_{C1} = 1) \\ i_{sto}(k+1) = -T_s u_{sto}(k)/L + i_{sto}(k), & (S_{C1} = 0) \end{cases}, \tag{8}$$

where $S_{C1} = 1$ represents the conducting state of switch V_1.

3.3.2. Objective Function of the Converter Predictive Control

In order to realize fast regulation of the inner loop current of the converter, predictive control should aim for fast tracking of the reference current generated from the outer loop, as shown in Equation (9).

$$J = \left| i_{sto}(k+1) - I^{*}_{sto} \right|. \tag{9}$$

On the basis of collecting the state information of the system at the current moment, the predicted values of the converter inductor current under different switching states can be calculated though the prediction model. The switch state with the smallest deviation between the predicted current and corresponding reference value is then selected as the system output to control the converter. As a result, fast current tracking can be achieved through active predictive control.

By combining the model predictive method of current inner loop with the adaptive droop control in the outer loop of the energy storage control system, fast power regulation of energy storage units under various working modes can be achieved, which stabilizes the DC voltage. The general control chart of the predictive method with adaptive droop control in the outer loop for different energy storage units is presented in Figure 6.

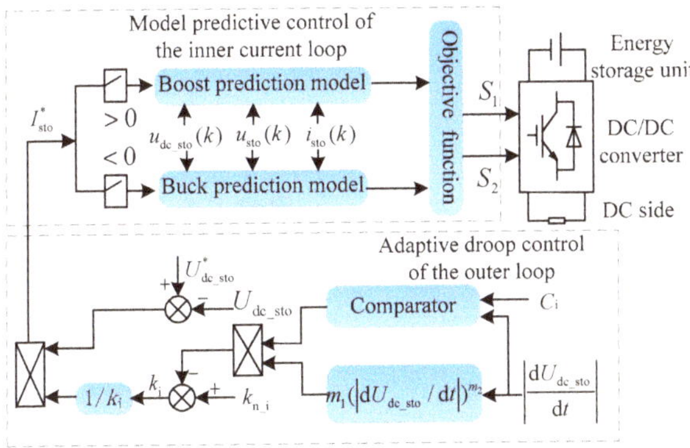

Figure 6. Predictive control model of an energy storage system adopting adaptive droop control in the outer loop.

As mentioned before, the supercapacitor is appropriate for suppressing high frequency power fluctuations with small amplitude, which are hard for the biomass thermal unit or the battery to stabilize. On the contrary, the battery, which has large amount of stored energy and high energy density, is always utilized as a long-term power balance device to absorb or release power with low frequency. The reference current absorbed or released by the energy storage system is divided into two parts, namely, the low-frequency part and the high-frequency one, through the low pass filter LPF. The corresponding control block diagram of the hybrid energy storage system is shown in Figure 7.

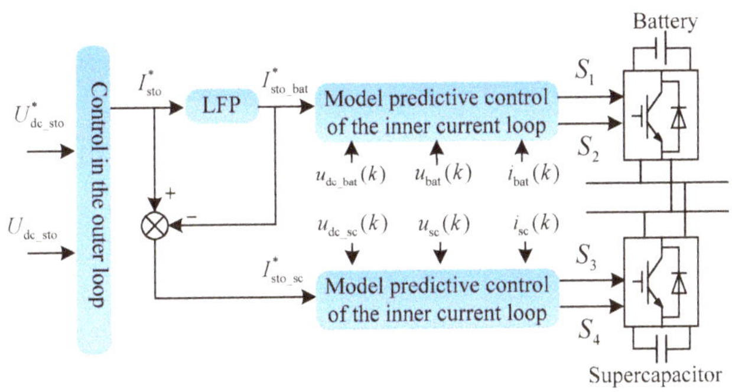

Figure 7. Control block diagram of the hybrid energy storage system.

3.4. Control of the Energy Stored in the Wind Turbine

In the wind power generation system, the wind turbine can store or release rotational kinetic energy through the change of its rotational speed, which makes it "the third energy storage unit" in the microgrid. However, in a converter-based microgrid, the direct connection between the disturbance in DC side and the change of rotational kinetic energy in the wind power system is isolated by the turbine-side converter. Generally, the wind power unit operates in maximum power point tracking (MPPT) mode, in which the output electric power is independent of the change of DC voltage. Therefore, auxiliary control is needed to establish the correlation between the kinetic energy of the wind turbine and the DC voltage, so that the mechanical rotation system can provide power support for the microgrid under conditions of significant change in DC voltage, to enhance its capability of fault ride-through.

When disturbance of DC bus voltage occurs, the charge or discharge power of the DC side capacitor in parallel with the wind power converter is calculated as follows.

$$\Delta P_{\mathrm{dc}} = C_{\mathrm{w}} U_{\mathrm{dc}} \frac{\mathrm{d}U_{\mathrm{dc}}}{\mathrm{d}t}, \tag{10}$$

where C_{w} is the DC side capacitor of the wind power system converter.

The variation of the output power caused by the change of the generator speed in the wind power unit is shown in Equation (11).

$$\Delta P_{\mathrm{e}} = \frac{\mathrm{d}E_{\mathrm{k}}}{\mathrm{d}t} = J\omega_{\mathrm{r}} \frac{\mathrm{d}\omega_{\mathrm{r}}}{p^2 \mathrm{d}t}, \tag{11}$$

where ΔP_{e} is the power variation of the wind power unit, E_{k} is the rotating kinetic energy of the generator, J is the rotating inertia of the synchronous generator, ω_{r} is the angular speed of the synchronous generator, and p is the number of pole pairs of the generator. As the wind turbine and the permanent magnet generator are directly coupled, the generator speed is equal to that of the wind turbine.

As the mechanical energy of the wind turbine is much larger than the energy stored in the capacitor, variation of the wind turbine speed is much smaller than that of the DC voltage when bearing the same amount of power disturbance. To quickly restrain the fluctuation of DC voltage with urgency, it is necessary to transform the power unbalance on the DC side into the change of rotational kinetic energy of the wind turbine through a control method. Therefore, fluctuation of DC bus voltage can be assumed by the wind turbine. The following equation is available by combining Equations (2) and (3).

$$J\omega_{\mathrm{r}} \frac{\mathrm{d}\omega_{\mathrm{r}}}{p^2 \mathrm{d}t} = C_{\mathrm{w}} U_{\mathrm{dc}} \frac{\mathrm{d}U_{\mathrm{dc}}}{\mathrm{d}t}. \tag{12}$$

The relationship between the variation of the wind turbine speed and the change of DC voltage under the same power disturbance is obtained by integrating and normalizing both sides of Equation (12) simultaneously. It is assumed that the voltage is kept at its rated value $U_{\mathrm{dc\_N}}$ before occurrence of the disturbance.

$$\omega_{\mathrm{r1\_pu}}^2 - \omega_{\mathrm{r0\_pu}}^2 = \frac{\frac{1}{2} C_{\mathrm{w}} U_{\mathrm{dc\_N}}^2}{\frac{1}{2p^2} J \omega_{\mathrm{r\_N}}^2} (U_{\mathrm{dc\_pu}}^2 - 1) = k_{reg} (U_{\mathrm{dc\_pu}}^2 - 1), \tag{13}$$

where $\omega_{\mathrm{r0\_pu}}$ and $\omega_{\mathrm{r1\_pu}}$ are the per-unit values of the angular speed of the generator before and after disturbance, respectively; $U_{\mathrm{dc\_pu}}$ is the per-unit value of DC voltage after disturbance; $\omega_{\mathrm{r\_N}}$ is the rated speed of the generator. k_{reg} is defined as the speed regulation coefficient, and varying degrees of regulation of the DC voltage can be achieved by setting different values of k_{reg} to ensure smooth transition of the microgrid during major disturbances.

Figure 8 is the switching principle of the improved MPPT curve with speed response of the wind turbine when disturbance occurs in the DC voltage of the microgrid. As shown in this figure, P_{opt} is the maximum power point tracking curve, which determines the output of wind power unit under normal conditions. In addition, the curves $P_{\mathrm{opt\_max}}$ and $P_{\mathrm{opt\_min}}$ are the upper and lower limits of the output power, respectively. Usually, the microgrid is stable and the wind power unit operates at point A, with the proportional coefficient of MPPT curve being k_{opt0}. When a steep fall of the DC voltage occurs due to a sudden power vacancy, a large amount of energy is needed to reduce the change rate of DC voltage so that load shedding can be avoided. At this time, the curve switches to $P_{\mathrm{opt\_max}}$ and the operating point moves from A to O, with the proportional coefficient rising to $k_{\mathrm{opt\_max}}$. The wind turbine, which connects directly to the generator, slows down due to the fact that the output electromagnetic power is larger than the mechanical power captured by the wind turbine, and the operation point decreases to B along $P_{\mathrm{opt\_max}}$. With the recovery of DC voltage, the proportional coefficient decreases gradually from $k_{\mathrm{opt\_max}}$ to k_{opt0}, with the power tracking curve cutting back to P_{opt} slowly and smoothly. Therefore, the wind turbine recovers and operates again at point A. Similarly, the switching process of the operation points can be analyzed when DC bus voltage surges.

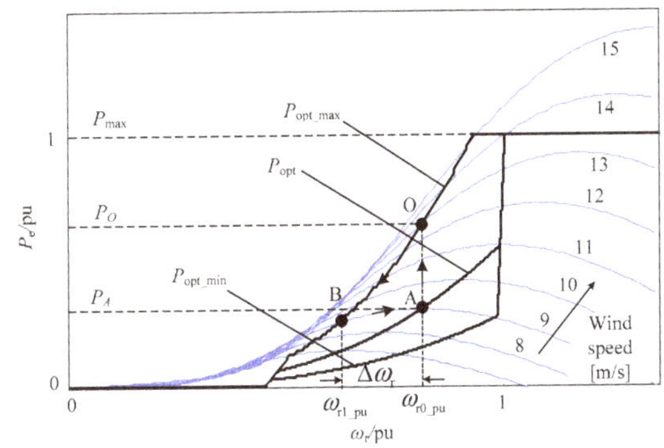

Figure 8. Switching principle of improved MPPT curve.

Point A and point B in Figure 8 correspond to the rotational speed ω_{r0_pu} and ω_{r1_pu}, respectively. When the range of speed regulation is not wide, the output power of point A and point B is approximately equal, as shown in Equation (14).

$$k_{opt}\omega_{r1\_pu}^{3} = k_{opt0}\omega_{r0\_pu}^{3}, \tag{14}$$

where k_{opt} is the proportion coefficient of power tracking curve after adopting the improved speed control. k_{opt} can be obtained by introducing Equation (13) into Equation (14).

$$k_{opt} = \frac{\omega_{r0\_pu}^{3}}{\left[\omega_{r0\_pu}^{2} + k_{reg}(U_{dc\_pu}^{2} - 1)\right]^{3/2}} k_{opt0}. \tag{15}$$

In order to adjust the inertia response of the wind power unit according to the change rate of DC voltage in microgrid, the speed regulation coefficient k_{reg} is modified by the change rate signal to be $k_{reg} = k_w\,|dU_{dc}/dt|$, where, k_w is a constant. By utilizing the proportional coefficient k_{opt} calculated in Equation (15) instead of the fixed proportional coefficient k_{opt0}, adaptive speed response of the wind power system can be achieved to quickly adjust the output power by using the energy stored in wind turbines. The block diagram of the improved control strategy for the wind power system is shown in Figure 9, where k_{opt_min} and k_{opt_max} are the limit values of the proportional coefficient.

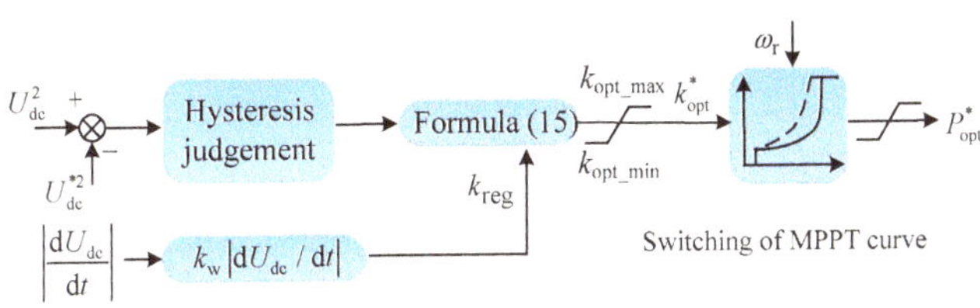

Figure 9. Principle diagram of self-adaptive speed control of wind power system.

4. Overall Control of the Forest Microgrid

The basic control and the improved control with hybrid complementary energy storage of the microgrid in different working modes are summarized in Figure 10. In free mode, the supercapacitor is supplementary to the main control of the biomass unit to smooth the DC voltage and enhance the power quality. In emergency mode, the battery and the supercapacitor take in charge of stabilizing high frequency and low frequency disturbances, respectively. In addition, as a special energy storage unit, the wind turbine can absorb superfluous energy when the DC voltage is higher than the upper limited value, and release its huge kinetic energy when the voltage decreases to the lower threshold to fill the power gap and avoid load shedding.

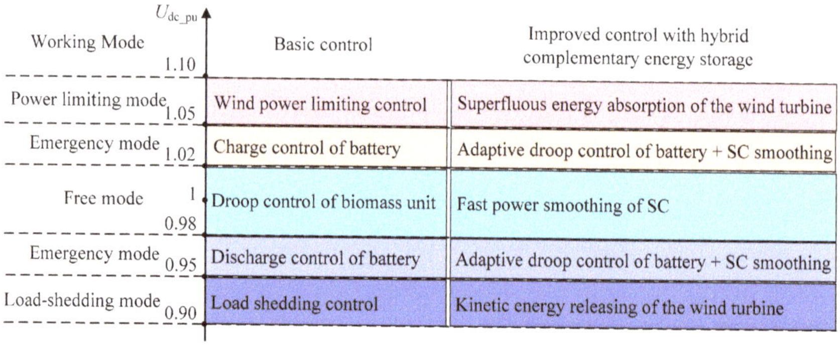

Figure 10. Comparison diagram of the basic control and the improved control of hybrid complementary energy storage.

As the control of the microgrid is based on the variation of DC voltage, operation mode of the system is primarily determined through the voltage hysteresis, as shown in Figure 11, where, u_{t1} and u_{t2} are the threshold voltages at mode switching points, which are selected to be 0.02 and 0.05, respectively, in this paper. Variables $S_0 = 1$, 2, and 3 indicate that the system is operating in free mode, emergency mode, and power limiting mode respectively. Moreover, voltage hysteresis control is adopted to avoid frequent switching of operation modes of the converters.

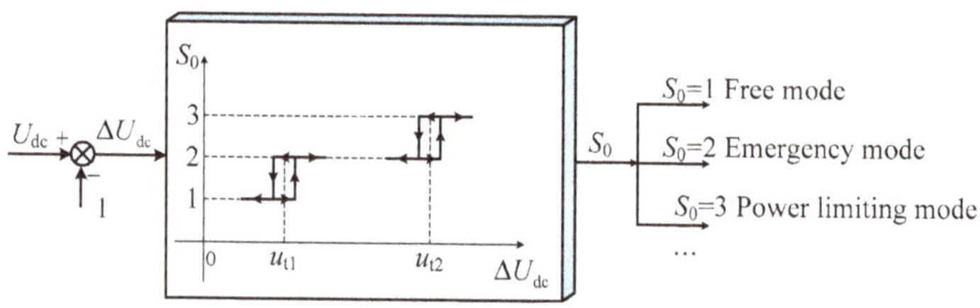

Figure 11. Operation mode judgment through voltage hysteresis.

The overall control structure of the islanded DC microgrid in forest area is presented in Figure 12, in which hybrid complementary energy storage is appreciated for power quality improvement and fault ride-through enhancement. S, S_1, S_2, and S_3 are the switches of the convertors for transferring among different operating modes.

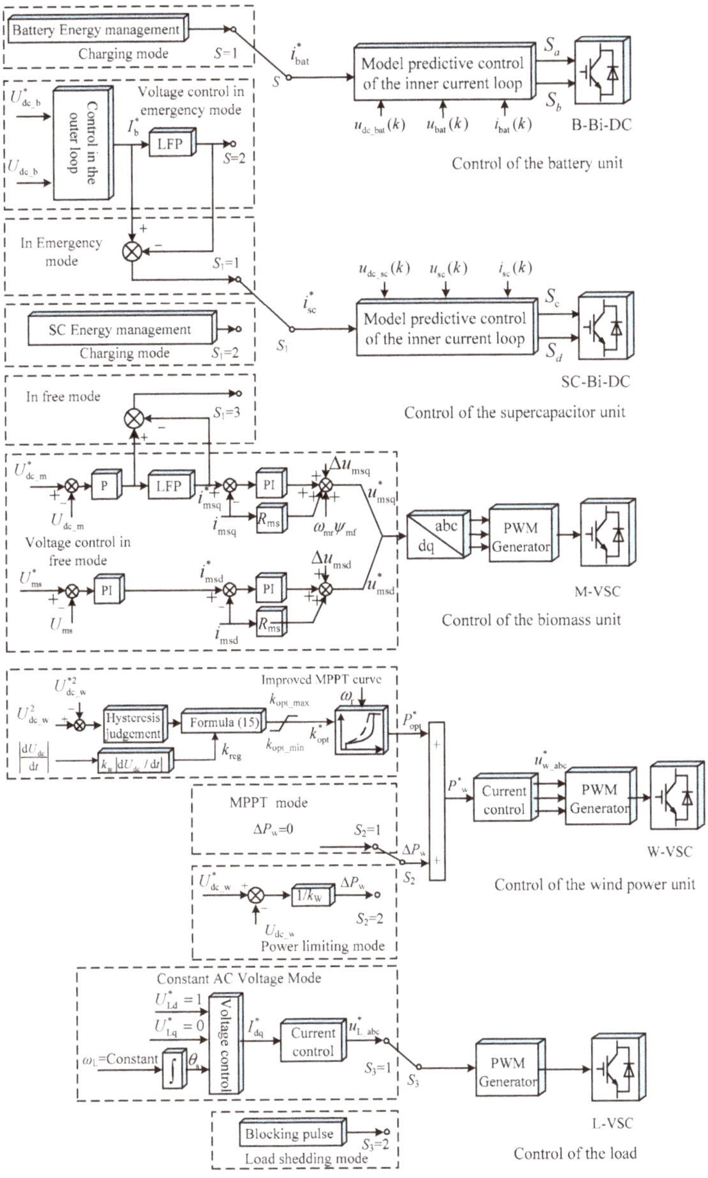

Figure 12. Overall control of the islanded DC microgrid in a forest area.

5. Simulation Analysis

5.1. Main Settings of the Simulations

In order to verify the effectiveness of the proposed sectional coordination control with hybrid complementary energy storage, a simulation model of the islanded DC microgrid in forest area was established in MATLAB/Simulink and simulations were conducted. The structure of the system is shown in Figure 1, in which the forest biomass power unit, the wind energy unit, and the energy storage units are connected to DC bus through electronic converters. The main parameters of the system are listed in Table 1. In the following, the effects of basic coordination control and improved control methods at both steady and transient states are presented and analyzed.

Table 1. System parameters.

Variable	Symbol	Value
Rated power of micro-gas turbine system	P_{mN}	65 kW
Rated voltage of micro-gas turbine	U_{mN}	400 V
Rated voltage of battery	U_{bN}	180 V
Rated capacity of battery	P_{bN}	100 A h
Rated capacity of battery converter	P_{b_DCN}	40 kW
Rated capacity of supercapacitor converter	P_{SCN}	40 kW
Rated speed of the wind turbine	ω_N	75 r/min
Number of wind turbine	N_w	2
Rated capacity of the wind turbine	P_{wN}	50 kW
Rated voltage of DC bus	U_{dcN}	400 V
Sampling period	T_s	50 µs

5.2. Simulation Analysis with Fault in the Biomass Unit

At the beginning of the simulation, the islanded forest microgrid operated steadily, with the wind power generation system, the hybrid energy storage system, and the biomass power unit providing stable power output to the DC side. At the 6th second of the simulation, the output of the biomass unit was reduced instantaneously due to a fault of the micro-gas turbine, which led to a large power gap in the DC microgrid. In Figure 13a, simulation results of dashed lines and solid lines are obtained when the basic method of sectional coordinated control and the improved method of hybrid complementary energy storage control based on the prediction model were adopted, respectively. It can be seen from the figures that when the basic method was utilized, the DC voltage dropped rapidly to about 0.93 pu, due to the sudden decline of the system power. According to the control principle of the converters in the DC microgrid, the system will switch to load-shedding mode and the load will be shed according to priority until the voltage restores. However, when the hybrid complementary energy storage control method based on converter prediction model was employed, once the system switched to emergency mode, the hybrid complementary energy storage unit started immediately. Under the fast predictive control of the converter, instantaneous power compensation was provided to relax the downtrend of DC voltage. At this time, the change of DC voltage did not exceed the mode switching threshold, which avoided load shedding and improved the capability of fault ride-through.

In order to compare the control speed of the predictive control method of converter with traditional PI control in the inner loop, the converter control with PI regulator in the inner loop of the hybrid energy storage was simulated and analyzed, as shown in the solid lines in Figure 13b. The dotted lines in the figure still indicate the simulation results with the basic coordinated control being used alone. At the instant of DC voltage drop, the output power of the hybrid energy storage system increased instantaneously, with adaptive modification of the adjustment coefficient in the outer loop of the control. However, the output power of the system could not meet the power requirement and presented unsatisfactory dynamic characteristics, due to the using of PI control which was essentially hysteretic in the inner loop. As a result, the DC voltage was still below the mode switching threshold, which made it difficult to achieve substantial improvement of power quality and reliability.

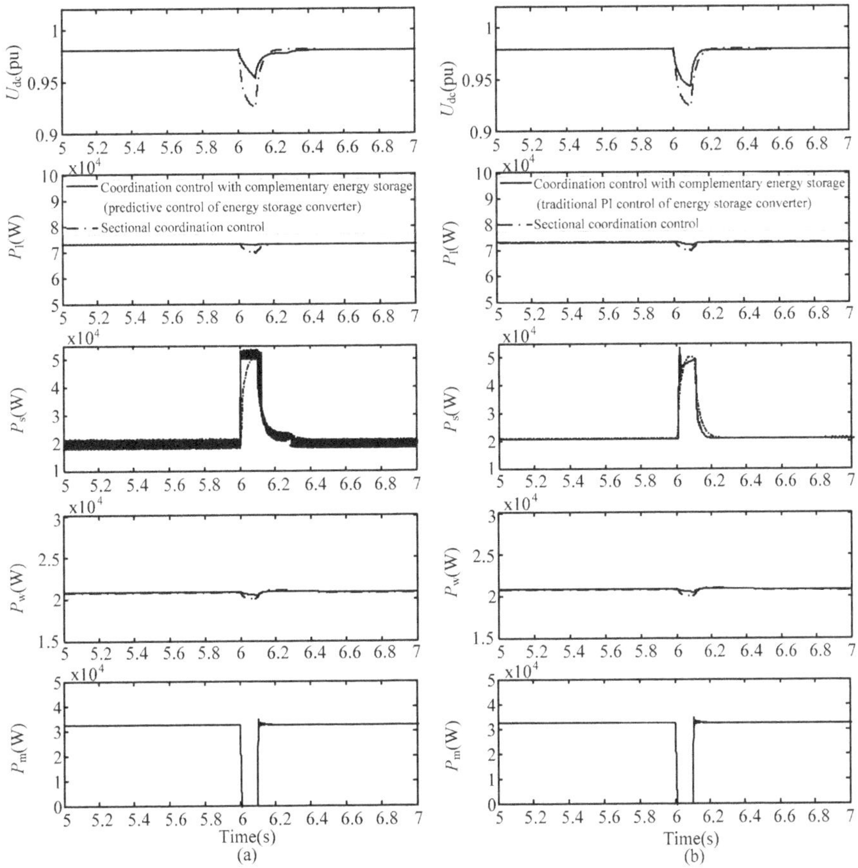

Figure 13. Performance of each unit in the DC microgrid under fault condition in the biomass unit: (**a**) The solid lines were obtained by using the improved method of hybrid complementary energy storage based on prediction and the dotted lines were obtained utilizing the basic coordinated control; (**b**) The solid lines were obtained by using the improved method based on PI control and the dotted lines were obtained utilizing the basic coordinated control.

5.3. Simulation Analysis with Fluctuating Loads

The islanded microgrid in forest is generally vulnerable to stochastic fluctuating loads, such as electrical machinery with characteristic of frequent start-up and shut-down, especially when the micro-gas turbine of the biomass unit has encountered fault or energy exhaustion. The following simulation presents the above extreme condition, with corresponding figures shown in Figure 14. In Figure 14a, the dashed lines and solid lines represent the simulation results when the basic method of sectional coordinated control and the improved method of hybrid complementary energy storage control based on prediction model were adopted, respectively. As is shown in the figures, when the basic method was utilized, although the energy storage system responded quickly, the fluctuation of DC voltage was large and the lowest point of voltage was below 0.95 pu. When the improved method was adopted, the hybrid energy storage system could rapidly increase the output power according to the voltage change rate, so as to provide rapid power support for the system. At the same time, the wind power system released its rotational kinetic energy and provided an intermittent energy supply to smooth the power peak. With the joint efforts of the battery, the capacitor, and the wind turbine, the lowest point of DC voltage was raised to about 0.97 pu, and the quality of the DC power supply was greatly improved.

When the hybrid energy control (with only battery and supercapacitor) based on traditional PI regulation was adopted, as shown in the solid lines in Figure 14b, the output power of the hybrid energy storage unit increased slightly compared with the basic control, and the corresponding fluctuation amplitude of DC voltage was between the value using the basic control and the one utilizing the proposed method.

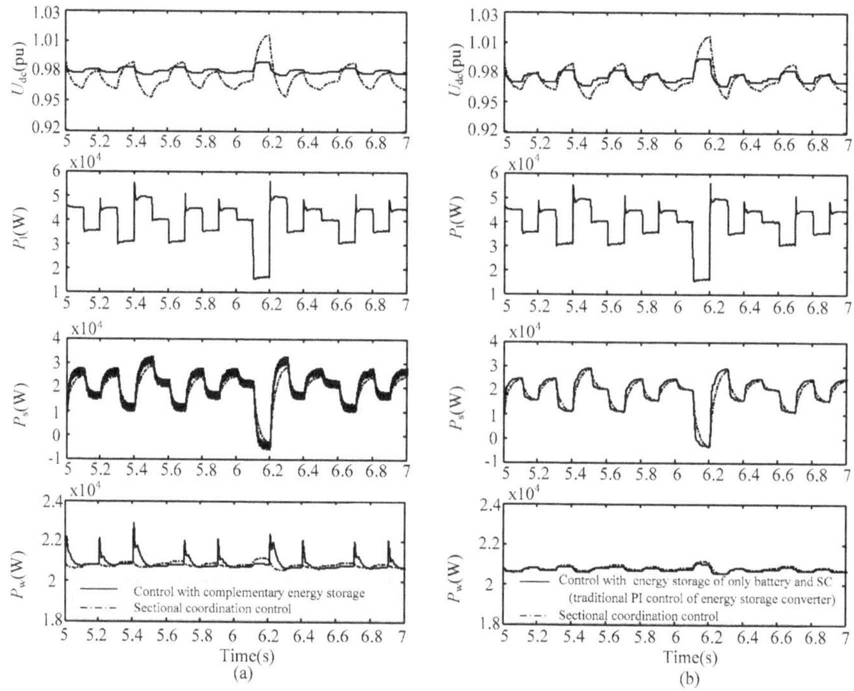

Figure 14. Performance of each unit in the DC microgrid with fluctuating load: (**a**) The solid lines were obtained by using the improved method of hybrid complementary energy storage based on prediction and the dotted lines were obtained utilizing the basic coordinated control; (**b**) The solid lines were obtained by using the improved method based on PI control and the dotted lines were obtained utilizing the basic coordinated control.

5.4. Simulation Analysis with Simulated Natural Wind Speed Similar to the Actual One

To simulate the actual wind speed, the superposition of basic wind, gust wind, and random wind was adopted as wind time series. The basic wind is a constant value, which was 7.5 m/s in this simulation. The gust wind v_z and the random wind v_n were obtained through the following Equation.

$$v_z = 0.5V_{zmax}\left[1 - \cos\frac{2\pi(t - T_I)}{T_g}\right], \tag{16}$$

$$v_n = V_{nmax} \cdot \text{unifrnd}(-1, 1) \cdot \cos[2\pi + \text{unifrnd}(0, 2\pi)], \tag{17}$$

where V_{zmax} and V_{nmax} represent the maxima of the gust wind and the random wind, respectively; T_I and T_g are the start time and the cycle of the gust wind, respectively; and unifrnd (a, b) is a function that generates uniformly distributed random numbers with a and b as upper and lower bounds.

As is shown in Figure 15, the output of wind power presented fluctuating characteristics. When basic coordination control was adopted, DC voltage was disturbed by the fluctuating power. Although the voltage could still be maintained above 0.98 pu, fluctuating voltage will adversely affect voltage-sensitive loads in forest areas, such as parameter detection devices, and reduce the quality of output power. When the hybrid energy storage method proposed in this paper was adopted, under fast predictive control of converter, the wind turbine unit and the energy storage systems provided corresponding power support for the system according to the change rate of DC voltage, with additional power output suppressing the random fluctuation. The proposed method ensured the stability of system power and improved the quality of the DC voltage.

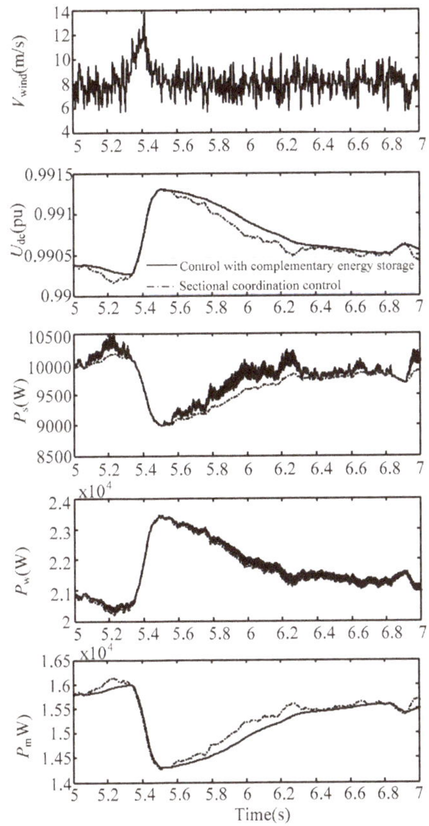

Figure 15. Performance of each unit in DC microgrid with simulated natural wind speed.

6. Conclusions

A hybrid complementary energy storage control method in forest DC microgrid is proposed, in which wind turbine energy storage, the battery, and the supercapacitor coordinate under different working conditions. Firstly, according to the characteristics of batteries and supercapacitors, batteries are adopted as a long-term energy reserve to bear the low-frequency fluctuation, while supercapacitors are used to suppress the corresponding high-frequency component. In addition, predictive control of the converters is adopted to reduce control delay and ensure the effectiveness of the energy storage converters. Furthermore, the inertia enhancement control strategy of the wind turbine system was studied by utilizing the large rotating kinetic energy of wind turbines to suppress power disturbance in a timely way and improve the fault ride-through capacity.

Through simulations of the DC microgrid in forest area conducted in MATLAB/Simulink, conclusions can be drawn that by utilizing the hybrid complementary energy storage, fast power smoothing can be achieved so that power quality is improved compared with methods using traditional PI control or fixed droop control. In addition, fault ride-through capability was effectively enhanced compared with microgrid with single energy storage by inertia control of the wind power unit that utilizes the kinetic energy in wind turbine.

Author Contributions: M.Y. proposed the algorithm; M.Y. and J.Z. conceived and designed the experiments; M.Y. and J.Z. performed the experiments; M.Y. and H.L. wrote the paper.

Funding: Project supported by funding for fundamental research business expenses of central universities (Grant No. BLX201716).

Conflicts of Interest: The authors declare no conflict of interest.

Nomenclature

$i_{bat}(k)$	Current of the battery at the *k*th moment
$i_{sc}(k)$	Current of the supercapacitor at the *k*th moment
P_w	Power generated by the wind turbine.
P_{sc}	Power of the supercapacitor.
P_b	Power of the battery.
P_m	Power generated by the biomass unit.
P_{li}	Power consumed by load i.
P^*_{opt}	Reference power of the wind power system
U_{dc}	DC voltage of the main DC bus
U_{dc_m}	DC voltage at the outlet of the biomass unit
U_{dc_w}	DC voltage at the outlet of the wind power unit
U_{dc_b}	DC voltage at the outlet of the battery unit
U_{dc_l}	DC voltage at the outlet of load
U_{dc_sto}	DC voltage at the outlet of energy storage unit
$U^*_{dc\_sto}$	Reference value of the DC voltage at the outlet of energy storage unit
$u_{bat}(k)$	Voltage of the battery at the *k*th moment
$u_{dc\_bat}(k)$	DC side voltage of the battery converter at the *k*th moment
$u_{sc}(k)$	Voltage of the supercapacitor at the *k*th moment
$u_{dc\_sc}(k)$	DC side voltage of the supercapacitor converter at the *k*th moment

References

1. Jodi, S.B.; Van, B.; Schwab, T.K.; Volker, C.R. The relative effectiveness of protected areas, a logging ban, and sacred areas for old-growth forest protection in southwest China. *Biol. Conserv.* **2015**, *181*, 1–8.
2. González, J.; Arnau, R.R.J.; Rius, C.A. Optimal sizing of a hybrid grid-connected photovoltaic–wind–biomass power system. *Appl. Energy* **2015**, *154*, 752–762. [CrossRef]
3. Teixeira, R.T.; Santos, A.R.D.; Marcatti, G.E. Forest biomass power plant installation scenarios. *Biomass. Bioenerg.* **2018**, *108*, 35–47. [CrossRef]
4. Soliño, M.; Prada, A.; Vázquez, M.X. Green electricity externalities: forest biomass in an Atlantic European Region. *Biomass. Bioenerg.* **2009**, *33*, 407–414. [CrossRef]
5. Martin, C.; Rocco, H.; Brandon, S. Climate change and power security: power load prediction for rural electrical microgrids using long short term memory and artificial neural networks. *Appl. Sci.* **2018**, *8*, 749.
6. Schonberger, J.; Duke, R.; Round, S.D. DC-bus signaling: a distributed control strategy for a hybrid renewable nanogrid. *IEEE Trans. Ind. Electron.* **2006**, *53*, 1453–1460. [CrossRef]
7. Peng, W.; Feng, C.; Che, A. Bi-Objective scheduling of fire engines for fighting forest fires: new optimization approaches. *IEEE Trans. Intell. Transp. Syst.* **2017**, *99*, 1–12.
8. Park, H.; Turner, N.; Higgs, E. Exploring the potential of food forestry to assist in ecological restoration in North America and beyond. *Restor. Ecol.* **2017**, *26*, 284–293. [CrossRef]

9. Kowalczyk, I. Barriers to the expansion of electrical cogeneration by the wood products industry in the United States. *J. Sustain. Forest.* **2013**, *32*, 159–174. [CrossRef]
10. Busarello, T.D.C.; Paredes, H.K.M.; Pomilio, J.A. Synergistic operation between battery energy storage and photovoltaic generator systems to assist management of microgrids. *IET. Gener. Transm. Dis.* **2018**, *12*, 2944–2951. [CrossRef]
11. Tan, M.; Cintuglu, M.H.; Mohammed, O.A. Control of a hybrid AC/DC microgrid involving energy storage and pulsed loads. *IEEE Trans. Ind. Appl.* **2017**, *53*, 567–575.
12. Karavas, C.S.; Arvanitis, K.G.; Kyriakarakos, G. A novel autonomous PV powered desalination system based on a DC microgrid concept incorporating short-term energy storage. *Sol. Energy* **2018**, *159*, 947–961. [CrossRef]
13. Sun, X.; Hao, Y.; Wu, Q. A multifunctional and wireless droop control for distributed energy storage units in islanded AC microgrid applications. *IEEE Trans. Power. Electr.* **2017**, *32*, 736–751. [CrossRef]
14. Korada, N.; Mishra, M.K. Grid adaptive power management strategy for an integrated microgrid with hybrid energy storage. *IEEE Trans. Ind. Electron.* **2017**, *64*, 2884–2892. [CrossRef]
15. Garcia-Torres, F.; Bordons, C. Optimal economical schedule of hydrogen-based microgrids with hybrid storage using model predictive control. *IEEE Trans. Ind. Electron.* **2015**, *62*, 5195–5207. [CrossRef]
16. Garcia-Torres, F.; Bordons, C.; Ridao, M.A. Optimal economic schedule for a network of microgrids with hybrid energy storage system using distributed model predictive control. *IEEE Trans. Ind. Electron.* **2019**, *66*, 1919–1929. [CrossRef]
17. Garcia-Torres, F.; Valverde, L.; Bordons, C. Optimal load sharing of hydrogen-based microgrids with hybrid storage using model-predictive control. *IEEE Trans. Ind. Electron.* **2016**, *63*, 4919–4928. [CrossRef]
18. Xiao, J.; Peng, W.; Setyawan, L. Hierarchical control of hybrid energy storage system in DC microgrids. *IEEE Trans. Ind. Electron.* **2015**, *62*, 4915–4924. [CrossRef]
19. Xu, Q.; Hu, X.; Peng, W. A decentralized dynamic power sharing strategy for hybrid energy Storage system in autonomous DC microgrid. *IEEE Trans. Ind. Electron.* **2017**, *64*, 5930–5941. [CrossRef]
20. Hosseinzadeh, M.; Salmasi, F.R. Robust optimal power management system for a hybrid AC/DC micro-grid. *IEEE Trans. Sustain. Energ.* **2015**, *6*, 675–687. [CrossRef]
21. Hosseinzadeh, M.; Salmasi, F.R. Fault-tolerant supervisory controller for a hybrid AC/DC micro-grid. *IEEE Trans. Smart Grid* **2018**, *9*, 2809–2823. [CrossRef]
22. Chang, X.; Li, Y.; Xuan, L. An active damping method based on a supercapacitor energy storage system to overcome the destabilizing effect of instantaneous constant power loads in DC microgrids. *IEEE Trans. Energy Conver.* **2017**, *32*, 36–47. [CrossRef]
23. Zhang, Y.; Yun, W.L. Energy management strategy for supercapacitor in droop-controlled DC microgrid using virtual impedance. *IEEE Trans. Power. Electr.* **2017**, *32*, 2704–2716. [CrossRef]
24. Awad, E.A.; Badran, E.A.; Youssef, F.H. Mitigation of switching overvoltages in microgrids based on SVC and supercapacitor. *IET. Gener. Transm. Dis.* **2017**, *12*, 355–362. [CrossRef]
25. Lu, X.; Sun, K.; Guerrero, J.M. Stability enhancement based on virtual impedance for DC microgrids with constant power loads. *IEEE Trans. Smart Grid* **2017**, *6*, 2770–2783. [CrossRef]
26. Bui, T.; Kim, C.H.; Kim, K.H.; Sang, B.R. A modular cell balancer based on multi-winding transformer and switched-capacitor circuits for a series-connected battery string in electric vehicles. *Appl. Sci.* **2018**, *8*, 1278. [CrossRef]
27. Worku, M.Y.; Abido, M.A.; Iravani, R. Power fluctuation minimization in grid connected photovoltaic using supercapacitor energy storage system. *J. Renew. Sustain. Energy* **2016**, *8*, 013501. [CrossRef]
28. Kim, S.K.; Chang, R.P.; Kim, J.S. A stabilizing model predictive controller for voltage regulation of a DC/DC boost converter. *IEEE Trans. Control Syst. Technol.* **2014**, *22*, 2016–2023. [CrossRef]

applied sciences

MDPI

Article

An AC/DC Distribution Network DG Planning Problem: A Genetic-Ant Colony Hybrid Algorithm Approach

Deyang Yin [1], Fei Mei [2] and Jianyong Zheng [1,*]

[1] School of Electrical Engineering, Southeast University, Nanjing 210096, China; 230179580@seu.edu.cn
[2] College of Energy and Electrical Engineering, Hohai University, Nanjing 211100, China; meifei@hhu.edu.cn
* Correspondence: jy_zheng@seu.edu.cn; Tel.: +86-1390-514-3113

Received: 13 January 2019; Accepted: 14 March 2019; Published: 22 March 2019

check for updates

Featured Application: The potential application of this paper is to propose a method to solve the planning problem of DG access to the AC/DC distribution network.

Abstract: The planning problem of distributed generators (DG) accessing the AC/DC distribution network is a hot research topic at present. In this paper, a location and volume model of DG is established that considers DG operation and maintenance costs, DG investment costs, system network loss costs, fuel costs, pollution compensation costs, and environmental protection subsidies. Furthermore, voltage and power constraints are also considered in the model. To solve the proposed model, a hybrid algorithm called the GA-ACO algorithm is presented that combines the ant colony algorithm (ACO) and the genetic algorithm (GA). On one hand GA has good robustness, good adaptability, and quick global searching ability but it also has some disadvantages such as premature convergence and low convergence speed. On the other hand, ACO has the ability of parallel processing and global searching but its convergence speed is very low at the beginning. The IEEE-33 node distribution network is taken as a basic network to verify the rationale of the proposed model and the effectiveness of the proposed hybrid algorithm. Simulation results show that the proposed model is very in line with reality, the hybrid algorithm is very effective in solving the model and it has advantages in both convergence speed and convergence results compared to ACO and GA.

Keywords: AC/DC distribution network; distributed generator; planning; timing characteristics; genetic-ant colony hybrid algorithm

1. Introduction

In today's society, the energy crisis and environmental protection problems have become more and more serious. Traditional fossil energy cannot meet the goals of sustainable human development. Distributed power generation technology based on renewable energy has attracted more and more attention. Distributed generators (DG) refers to a small generator set that is designed, installed, and operated in a distribution network with capacities ranging from a few kilowatts to tens of megawatts [1]. Due to its high reliability, and clean, environmentally friendly, and flexible installation location, DG plays an increasingly important role in the distribution network.

With the promotion and popularization of DG and the stricter requirements of users for power quality, the traditional AC distribution network has revealed its inability to accept new energy [2–4]. In recent years, the development of power electronic devices has made great progress. The continuous improvement of converter devices has accelerated the research on related technologies of DC distribution networks. Compared with AC distribution networks, DC distribution networks have a larger power capacity, and the power loss is reduced, thus, the power quality becomes higher, and the distributed power source has easier access. This has made it more and more popular with scholars. If a DC distribution network is used, the converter used for DG access is saved greatly, and the energy loss is reduced; also, the DC bus has no phase and frequency synchronization problems, making the control of the distributed power supply simple and reliable [5]. Therefore, there is an urgent need to study the problem of DG access to AC /DC distribution networks.

When the DG is connected to the distribution network, the direction of the system power changes, which causes changes in the distribution network loss, so that the network loss is not only related to the load, but also related to the location and quantity of the DG. At the same time, due to the intermittent nature, volatility and randomness of DG, it inevitably affects the safe, stable and reliable operation of the distribution network. If the penetration rate of the distributed power supply is too high or the location of the access distribution network is improperly selected, it will not improve the environmental protection and economy of the grid operation, but will affect the safe and stable operation of the system. Therefore, it is necessary to plan the construction of the distribution network with DG.

At present, the problem of distribution network planning with DG has been studied from different perspectives. El-Khattam et al. [6] compared the difference between the distribution network with DG planning and traditional distribution network planning, and elaborated the DG planning problem from the aspects of economy and reliability. Zeng et al. [7] adopted the two-layer scene planning to solve the optimal distribution network planning scheme. Li et al. [8] established the objective function by reducing the network loss and improving the power quality, and adopted an intelligent algorithm to solve the DG location and capacity determined problem. Zhu D et al. [9] mainly considered DG and utilized the uncertainty of load growth to establish the distribution network planning model with DG and grid as the planning object. Guo et al. [10] comprehensively considered the stochastic volatility of DG and load, and used the opportunity constrained programming method to comprehensively optimize the configuration of DG. Rau N S et al. [11] used the gradient method to find the optimal solution of DG installation and configuration problems, but found it was easy to fall into local parts. Wang et al. [12] considered the location and capacity of DG and the expansion planning of the distribution network, and used a combination of genetic algorithm, branch exchange and simulated annealing algorithm to solve the model. Mei et al. [13] used the improved particle swarm optimization algorithm with simulated annealing algorithm to solve the problem of DG site selection and volume calculation based on the optimal network loss. Ye et al. [14] used the adaptive mutation particle swarm optimization algorithm to plan the location and volume of DG without considering the load-added nodes. However, at this stage, distribution network planning with DG mainly stays in the AC power distribution stage, and there is very limited research content on AC/DC distribution network planning. At the same time, most of the literature on the planning process is mainly based on the rated capacity of DG, and does not take the timing characteristics of the DG and the load into account. This causes deviations in the actual planning problem, and it is necessary to fully consider the DG and load timing characteristics in the planning process.

In recent years, more and more research including the application of heuristic intelligent optimization algorithms has been carried out. Genetic algorithms (GA) and ant colony algorithms (ACO) serve as two mainstream algorithms, and each of them has unique advantages. The genetic algorithm draws on the biological evolutionary law of nature to optimize the solution of the problem by simulating the evolution process of biological genes. The ant colony algorithm mimics real-world ant colony behavior. The algorithm simulates group behavior composed of simple individuals and seeks optimal results through group behavior.

Therefore, this paper establishes a DG access AC/DC distribution network planning model that takes environmental costs and timing characteristics into account, and optimizes the type, location and capacity of DG in the distribution network. At the same time, this paper studies the ideas and methods of GA and ACO to solve the problem, and analyzes the solution process. Then, by combining and improving the two algorithms, a GA-ACO algorithm is proposed and applied to the model solution.

2. AC/DC Distribution Network Model

The AC/DC distribution system generally consists of three main components: an AC distribution network, a DC distribution network, and a voltage source converter (VSC).

The main function of the VSC is to realize the bi-directional flow of active power on the AC side and the DC side, and at the same time to regulate the reactive power. The control mode of the converter mainly includes master-slave control and droop control [15–17]. This paper mainly studies the distribution network DG planning method when the converter adopts master-slave control mode.

To study the AC/DC distribution network trend, we must first establish a VSC model. When the VSC model is obtained, the power flow additional equation of the DC system should be given to consider the different control modes of the converter, so that the power flow of the AC/DC power distribution system can be solved. In the AC/DC distribution network described in this paper, the quasi-steady state model is used to equivalently process the DC part. The model not only effectively reflects the power characteristics of the DC converter, but also accurately and fully meets the actual engineering needs. The specific modeling process is as follow [18].

2.1. AC Distribution Network Power Flow Model

$$P_{jk} = P_{ij} - \frac{r_{ij}[(P_{ij})^2 + (Q_{ij})^2]}{(V_i)^2} - P_j \quad (1)$$

$$Q_{jk} = Q_{ij} - \frac{x_{ij}[(P_{ij})^2 + (Q_{ij})^2]}{(V_i)^2} - Q_j \quad (2)$$

$$(V_j)^2 = (V_i)^2 - 2(r_{ij}P_{ij} + x_{ij}Q_{ij}) + \frac{[(r_{ij})^2 + (x_{ij})^2][(P_{ij})^2 + (Q_{ij})^2]}{(V_i)^2} \quad (3)$$

$$P_j = P_{j,L} - P_{j,\mathrm{DG}} \quad (4)$$

$$Q_j = Q_{j,L} - Q_{j,\mathrm{DG}} \quad (5)$$

$$Q_{j,\mathrm{DG}} = P_{j,\mathrm{DG}} \tan\varphi \quad (6)$$

where V_i and V_j are the voltage amplitudes of node i and node j respectively; r_{ij} and x_{ij} are the branch resistance and reactance between node i and node j, respectively; P_{ij} and Q_{ij} are the active and reactive power of branch ij, respectively; $P_{j,L}$, $P_{j,\mathrm{DG}}$, $Q_{j,L}$, and $Q_{j,\mathrm{DG}}$ are the load active power, DG active power, load reactive power, and DG reactive power at node j, respectively. φ is the power factor angle.

2.2. DC Distribution Network Power Flow Model

According to the branch flow model of the AC distribution network, the branch flow model of the DC distribution network can be derived as follows:

$$P_{\mathrm{dc},jk} = P_{\mathrm{dc},ij} - \frac{r_{\mathrm{dc},ij}(P_{\mathrm{dc},ij})^2}{(V_{\mathrm{dc},i})^2} - P_{\mathrm{dc},j} \quad (7)$$

$$(V_{\mathrm{dc},j})^2 = (V_{\mathrm{dc},i})^2 - \frac{r_{\mathrm{dc},ij}P_{\mathrm{dc},ij}}{V_{\mathrm{dc},i}} \quad (8)$$

$$P_{\mathrm{dc},j} = P_{\mathrm{dc},j,L} - P_{\mathrm{dc},j,DG} \quad (9)$$

where $V_{\mathrm{dc},i}$ is the voltage amplitude of the DC node i; $r_{\mathrm{dc},ij}$ is the resistance between the branches ij; $P_{\mathrm{dc},ij}$ is the active power between the branches ij; $P_{\mathrm{dc},j,L}$ and $P_{\mathrm{dc},j,\mathrm{DG}}$ are the load active power and

DG active power at the node j, respectively. The VSC model is shown in Figure 1. Where $V\angle\theta_V$ is the voltage at the junction of the converter and the AC distribution system, R_{VSC} and X_{VSC} are the equivalent resistance and reactance inside the converter, $V_{VSC}\angle\theta_{VSC}$ is the phase voltage of the input converter, $P_{VSC} + jQ_{VSC}$ is the power of the input converter, P_{dc} and V_{dc} are the output power and voltage of the converter.

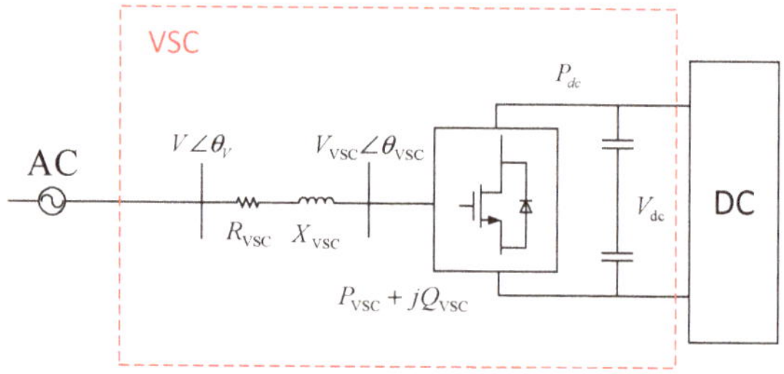

Figure 1. Voltage source converter (VSC) model.

The branch flow model of $R_{VSC} + jX_{VSC}$ is as shown in Equation (10). The AC three-phase active power of the input VSC is equal to the DC power of the VSC output. The input voltage and the output voltage satisfy the following relationship:

$$V_{VSC} = \frac{\sqrt{3}}{3}\mu V_{dc} \tag{10}$$

where μ is the DC voltage utilization under SPWM modulation, which is 0.866.

3. DG and Load Timing Characteristics

3.1. DG Timing Characteristics

Distributed generator usually contains micro-turbine generators (MT), wind turbine generators (WGs), photovoltaics (PV), and so on. Different types of DG apply to different regions. For example, photovoltaics generate more electricity in areas with sufficient light; micro-turbine generators are more suitable for areas with high heat demand [19]. Different DGs contribute differently to the environment, PV and WG are cleaner distributed generators. However, the timing fluctuation characteristics of these two generators, increase the uncertainty and affect the stability of the system while improving environmental protection. We have fully considered the timing characteristics of PV and WG. If we do not, there is no essential difference between WG and PV, and we cannot reflect on the advantages of MT. Additionally, we do not need to install energy storage equipment in the distribution network as this is contrary to the reality. Therefore, for the DG planning problem, the timing characteristics must be taken into consideration.

We took the three types of DG mentioned above as the research objects because these three types of DG are highly representative: the MT has rated capacity; its output is controllable and does not change with time; and the PV and WG represent a type of DG whose capacity changes with time. We also selected these two types of DG to study the impact of timing characteristics on the distribution network planning in depth.

Wind speed has a great influence on the WG. Generally, wind speed is high in the evening and especially in the winter. Therefore, the output is highest in winter and the least in summer. Light intensity and temperature have a great impact on PV. During the day, solar energy is rich, and the PV output is strongest at noon. At night, PV does not generate electricity. In general, PV has the largest output in summer and the least in winter. PV and WG have a natural complementarity, which is the main reason for selecting these two DGs which have different timing characteristics. According to meteorological data [20], wind speed curves and light intensity were obtained in different seasons, and the PV and WG timing characteristics curves were obtained as shown in Figures 2 and 3.

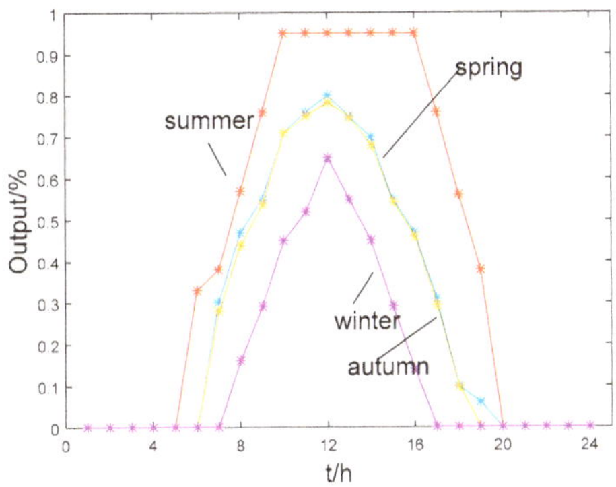

Figure 2. Photovoltaics (PV) timing characteristics curve.

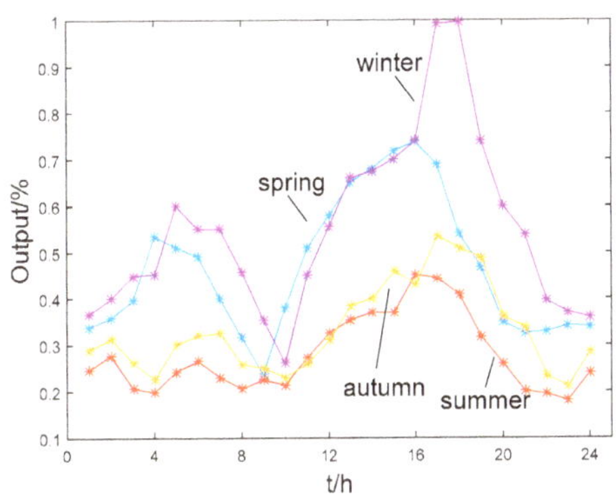

Figure 3. Wind turbine generators (WG) timing characteristics curve.

3.2. Load Timing Characteristics

At present, according to power planning and power industry statistics, the power load is generally divided into four typical loads, including industry, commerce, agriculture, and residents. Similar to DG, these loads also have particular timing characteristics. Moreover, the maximum output time of the DG is not always the same as the time of the maximum load. We took residential load and commercial load as the research objects of the distribution network, and their four seasonal load curves are shown in Figures 4 and 5. The electricity consumption of residents is the lowest in the summer. The daily peak hours are generally at noon and in the evening. The daily maximum load time is about 20:00. The minimum load time, except in summer, is about 03:00, while the summer minimum load time is about 07:00 in the morning. The fluctuations in commercial electricity consumption over the four seasons are small, the power consumption period ranges from 09:00 to 23:00, and the maximum load time is about 10:00.

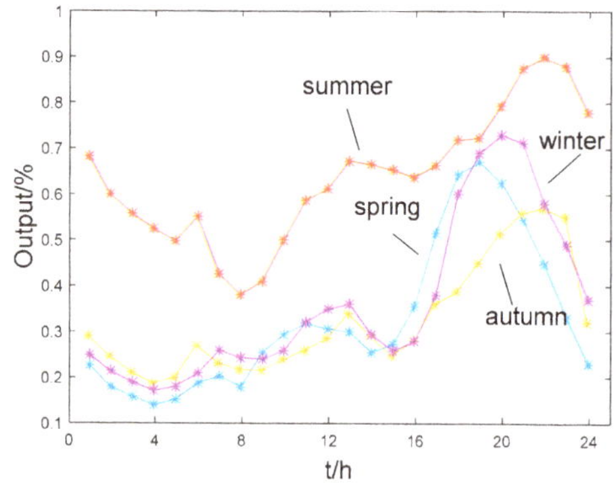

Figure 4. Residential loads timing characteristics curve.

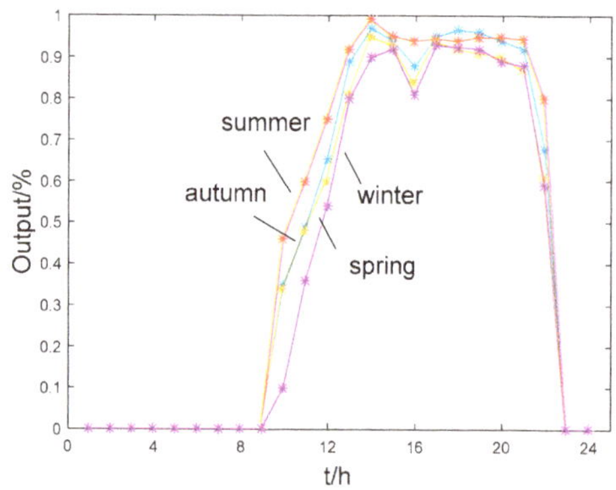

Figure 5. Commercial loads timing characteristics curve.

4. DG Optimization Configuration Model of Distribution Network

The objective function of this paper is to minimize the total annual system cost. Based on the distributed power supply capacity and voltage fluctuation range allowed by the distribution network node, the location and capacity of the DG are finally determined.

4.1. Objective Function

The annual cost of the distribution network is

$$\min C_{TOL} = C_{om} + C_i + C_{loss} + C_f + C_e - C_p \tag{11}$$

where C_{TOL} is the total annual cost of distribution system; C_{om} is DG annual operation and maintenance costs; C_i is the DG annual equivalent investment cost; C_{loss} is the system network loss cost; C_f is the MT fuel cost; C_e is the pollution compensation cost; and C_p is the environmental protection subsidy fee.

(1) DG annual operation and maintenance costs

$$C_{om} = c_{om} \sum_{t=1}^{8760} \sum_{i \in N_{DG}} E_i(t) \tag{12}$$

where c_{om} is DG unit operation and maintenance costs, 10,000 yuan/(MW·h); and $E_i(t)$ is the DG power generation connected to the i-th node at time t, MW·h.

(2) DG annual equivalent investment cost

$$C_i = \frac{r(1+r)^{n_y}}{(1+r)^{n_y}-1}\sum_{i\in N_{DG}}(c_i S[i]) \tag{13}$$

where r is the discount rate; n_y is the DG payback period; c_i is the investment cost of installing DG at the i-th node, 10,000 yuan; and $S[i]$ is the number of DG installations at the i-th candidate node.

(3) System network loss cost

$$C_{loss} = c_{loss}\sum_{t=1}^{8760}\sum_{i\in T} I_i^2(t) r_i \tag{14}$$

where c_{loss} is the unit network loss; I_i is the current on the i-th line, kA; and r_i is the resistance of the i-th line, Ω.

(4) MT fuel cost

$$C_f = c_f\sum_{t=1}^{8760}\sum_{i\in N_{mt}} E_{mt,i}(t) \tag{15}$$

where c_f is the unit fuel cost per hour, 10,000 yuan/(MW·h); and $E_{mt,i}(t)$ is the MT power generation connected to the i-th node at time t, MW·h.

(5) Pollution compensation cost

$$C_e = K_{mt}\sum_{i\in N_{mt}} E_{mt,i}(V_{CO_2}+R_{CO_2}) * 10^{-4} \tag{16}$$

where K_{mt} is the emission intensity of greenhouse gas produced by a unit MT, kg/(MW·h); V_{CO_2} is the environmental protection value, yuan/kg; and R_{CO_2} is a fine for the MT to emit greenhouse gases, yuan/kg.

(6) Environmental protection subsidy fee

$$C_p = \sum_{t=1}^{8760}\Big(\sum_{i\in N_{pv}} c_{pv}E_{pv,i}(t) + \sum_{i\in N_{wg}} c_{wg}E_{wg,i}(t)\Big) \tag{17}$$

where c_{pv} and c_{wg} are respectively PV and WG environmental subsidies for unit power generation, 10,000 yuan/(MW·h); $E_{pv,i}(t)$ and $E_{wg,i}(t)$ are respectively the PV and WG power generations connected to the i-th node at time t, MW·h.

4.2. Constraints

(1) DG installation capacity constraint

Light intensity, temperature, wind speed are all uncertain factors. If too much of this type of DG is added to the distribution network, the power quality of the system will be degraded. It is therefore necessary to limit the installation capacity of PV and WG.

$$\begin{cases} S_{pv,i} \le S_{pv,max}, i\in N_{pv} \\ S_{wg,i} \le S_{wg,max}, i\in N_{wg} \end{cases} \tag{18}$$

where $S_{pv,i}$ and $S_{wg,i}$ are respectively the PV installation capacity at the i-th PV candidate node and the WG installation capacity at the i-th WG candidate node; $S_{pv,max}$ and $S_{wg,max}$ are respectively the maximum installation capacity of PV and WG.

(2) Power flow equation

$$\begin{cases} P_i = U_i \sum\limits_{j \in i} U_j (G_{ij} \cos\theta_{ij} + B_{ij} \sin\theta_{ij}) \\ Q_i = U_i \sum\limits_{j \in i} U_j (G_{ij} \cos\theta_{ij} - B_{ij} \sin\theta_{ij}) \end{cases} \quad (19)$$

where P_i and Q_i are the active power and reactive power injected into node i, respectively; U_i is the voltage amplitude of node i, $j \in i$ represents all nodes connected to node i; G_{ij} and B_{ij} are respectively the admittance matrix real part and virtual part of the system; and θ_{ij} is the phase angle difference between node i and node j.

(3) Voltage constraint

$$U_{i,\min} \le U_i \le U_{i,\max} \quad (20)$$

$$U_{\mathrm{dc},i,\min} \le U_i \le U_{\mathrm{dc},i,\max} \quad (21)$$

where $U_{i,\max}$ and $U_{i,\min}$ are the upper and lower limits of the AC node voltage amplitude; $U_{\mathrm{dc},i,\max}$ and $U_{\mathrm{dc},i,\min}$ are the upper and lower limits of the DC node voltage amplitude.

(4) Branch flow constraint

$$S_j \le S_{j,\max} \quad (22)$$

$$S_{\mathrm{dc},j} \le S_{\mathrm{dc},j,\max} \quad (23)$$

$$S_{\mathrm{VSC},j} \le S_{\mathrm{VSC},j,\max} \quad (24)$$

where S_j is the power on branch j; $S_{j,\max}$ is the maximum allowable capacity on AC branch j; $S_{\mathrm{dc},j,\max}$ is the maximum allowable capacity on DC branch j; $S_{\mathrm{VSC},j,\max}$ is the maximum allowable capacity on VSC.

(5) Pollution compensation cost

$$\sum_{i \in N_{DG}} E_i(t) \ge L_{tol}(t) \ \mathrm{t} = 1,\ 2, \ldots 8760 \quad (25)$$

where $\sum\limits_{i \in N_{DG}} E_i(t)$ is the total capacities of the DGs at the time t, and $L_{tol}(t)$ is the sum of the loads of all nodes at time t.

5. Model Solving

In this paper, a GA-ACO algorithm is used to solve the model. The genetic algorithm mimics the generation and evolution process of all life intelligences. Using group search technology, through the genetic manipulation of selection, crossover and mutation of the current population, new groups are generated, and the population is gradually evolved to a state containing, or close to, the optimal solution. The ant colony algorithm searches for optimal solutions by simulating the processes that ant colonies use to search for food. Ants leave pheromones on the path they walk. The more ants that pass along the path, the more pheromones they leave, and the greater the probability that later ants will choose the path, thus forming a positive feedback mechanism, and finding the optimal solution.

The genetic algorithm has a large-scale and diverse initial population, so that global optimization is performed at the beginning of the algorithm, and the genetic algorithm is scalable and easy to combine with other algorithms. However, in the late stage of the genetic algorithm, the feedback information cannot be easily used, and a large number of redundant iterations are generated, which affects the convergence speed and accuracy of the algorithm. The ant colony algorithm has a fast convergence rate and good global convergence. In particular, when the pheromone is accumulated to a certain extent, the algorithm can quickly find the optimal solution. However, the disadvantage is that the accumulation of pheromones is slow at first, and it is easy to fall into the local optimum at the beginning.

Based on the advantages and disadvantages of the above two algorithms, the two algorithms can be combined to make up for their respective shortcomings. In this paper, the genetic algorithm is used to generate the initial solution, then these initial solutions are transformed into the pheromone distribution in the ant colony algorithm. Finally, the positive feedback mechanism of the ant colony algorithm is used to search for the optimal solution. The GA-ACO algorithm solution flow chart is shown in Figure 6.

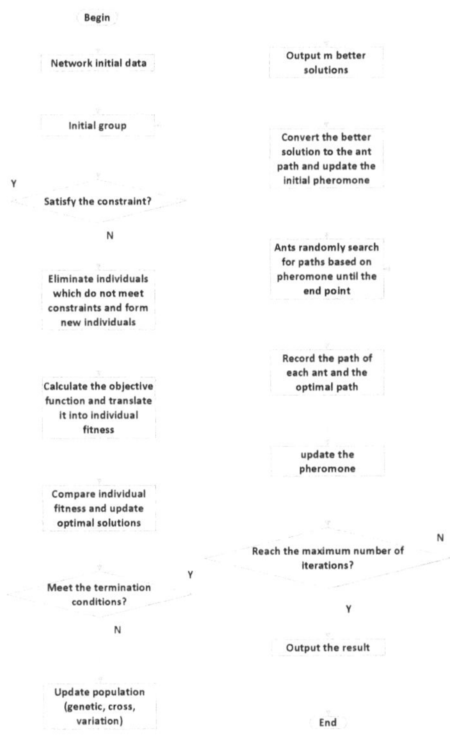

Figure 6. Algorithm flow chart.

5.1. *Genetic Algorithm Solving*

In this paper, we use binary coding. Each chromosome is regarded as a planning scheme. Each chromosome contains N_{DG} elements. The first N_{MT} elements represent the installed number of MT. The intermediate Npv elements represent the number of PV installations at each PV node to be selected; the last Nwg elements represent the number of WG installed at each WG node to be selected.

(1) Initial Population

The initial population of the genetic algorithm is randomly generated. To solve this model, a certain chromosome in the initial population does not meet the constraints. Combined with the planning model, the initial population is required to fully satisfy the system capacity constraints. Therefore, it is necessary to carry out innate elimination, and the individuals who do not satisfy the constraints are eliminated and regenerated until the number of individuals satisfy the constraints and reach the requirements of the initial population.

(2) Selection Operator

According to the fitness value, the individuals in the population are selected to obtain the parent sample. This paper uses the optimal individual preservation and roulette strategy to select operators. That is, the individual with the highest fitness is saved as a parent. For the remaining individuals, the roulette method is used to calculate the selected probability of each individual's fitness to select the retained individual. That is, the more it satisfies the target condition, the easier it is to be inherited by the offspring, but other individuals have offspring with a small probability to prevent the algorithm from falling into local optimum and guarantee global convergence.

(3) Crossover Operator

For individuals in the population, cross operations are performed according to a certain crossover probability, and corresponding mutation operations are performed according to the probability of a certain mutation generating a next generation population. In this paper, the crossover operation uses a two-point crossover operator, and the mutation operation uses a single-point mutation operator.

(4) Eliminate

Because the genetic algorithm generates the randomness of the individual, the generated offspring may not satisfy the constraint. This paper sets a penalty function. When the mismatching of the offspring is higher, the degree of the penalty is larger, making the gene of the offspring difficult for the next generation to inherit, thus ensuring the innate superiority of the population.

5.2. Ant Colony Algorithm Solving

(1) Pheromone Initialization

According to the location selection of the DGs, assuming that there are N installable DGs nodes, and the DGs capacity that may be installed on the node i is n_i, then a similar matrix as shown in Figure 7 can be generated.

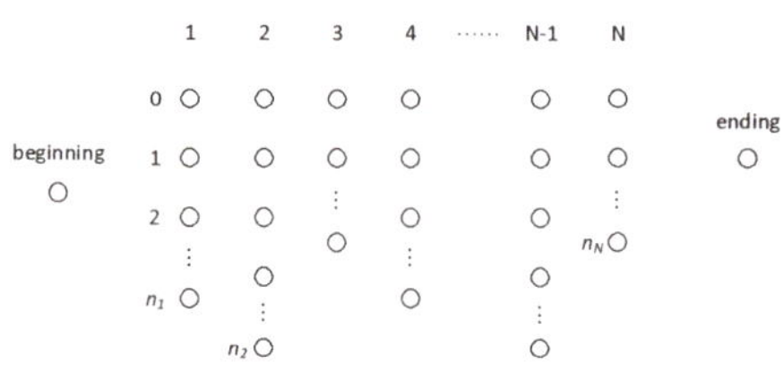

Figure 7. Ant colony algorithm search path diagram.

After T generation genetic optimization by genetic algorithm, m better solutions are obtained. Then the m preferred solutions are converted into the location and capacity of the ant colony algorithm, as shown in Figure 8. These location capacities are connected to form m paths, and update the pheromones on those paths. The number of ants is also set to m. M slaves are placed on m paths and the previous update pheromone are used as the initial value of the pheromone.

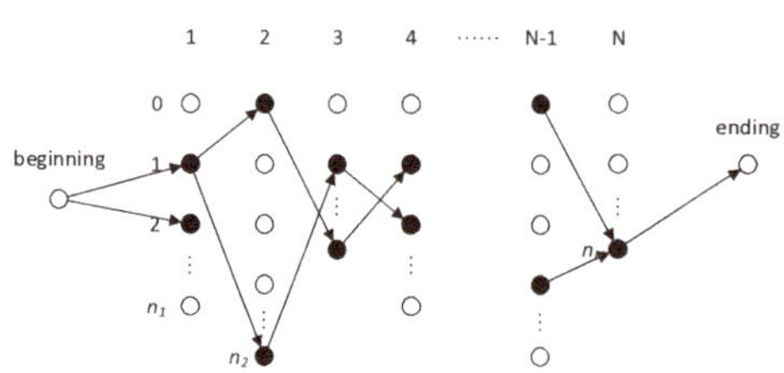

Figure 8. Genetic algorithm result conversion.

(2) Loop Iteration

In each iteration, ant k (k = 1, 2, 3, ..., m) determines the direction of their transition based on the pheromone on each path. The tabu table, $tabu_k$ is used to record the location capacity of the ants. The ant determines the transition probability based on the pheromone and path heuristic information on each path. $P_{ij}(t)$ represents the probability that ant k is transferred from position i to position j at time t:

$$p_{ij}^k = \begin{cases} \frac{[\tau_{ij}(t)]^\alpha [\eta_{ij}]^\beta}{\sum\limits_{s \in allowed_k} [\tau_{is}(t)]^\alpha [\eta_{is}]^\beta} & j \in allowed_k \\ 0 & \end{cases} \tag{26}$$

where $allowed_k = \{C - tabu_k\}$ (k = 1, 2, 3, ..., m) represents the position that the ant k next allows to select. α is a heuristic information factor, and represents the importance of the motion trajectory, that denotes the role that information accumulated by ants during exercise plays in the selection of ant movements. α has a value range of (0,5). β is the heuristic factor of expectation, which is the importance of characterizing the visibility of the path, that is, the role that the ant plays in the ant selection path during the movement. The value range of β is (0,5).

$\tau_{ij}(t)$ represents the pheromone strength on the path (i, j) at time t. Where i is the beginning location and j is the end location. $\eta_{ij}(t)$ is a heuristic function, which is generally the reciprocal of the sum of the DG installation cost at location j and the distributed power source operating cost.

(3) Pheromone Update

According to the expected value of each capacity of each node and the pheromone concentration value on the path, the transition probability of the n-th ant at each capacity point on the node is calculated, and then the path is selected.

In order to prevent excessive heuristic pheromone and flooding the heuristic information, the residual pheromone is updated after the ant completes an ergodic operation on all m position capacities. According to Equations (27) and (28), adjusting the amount of information on the path (i, j) at time t+1

$$\tau_{ij}(t+1) = (1-\rho)\tau_{ij}(t) + \Delta\tau_{ij}(t) \tag{27}$$

$$\Delta\tau_{ij}(t) = \sum_{k=1}^{m} \Delta\tau_{ij}^k(t) \tag{28}$$

where ρ is the pheromone volatilization coefficient, $1-\rho$ is the pheromone residual coefficient, and ρ is in the range of (0, 1). $\Delta\tau_{ij}(t)$ represents the pheromone increment on the path (i, j) during the current cycle, $\Delta\tau_{ij}(t)$=0 at the initial time. $\Delta\tau_{ij}^k(t)$ represents the number of pheromones left by the k-th ant in the path (i, j).

$$\Delta\tau_{ij}^k(t) = \begin{cases} Q/L_k & \text{if the kth ant passes the path (i,j)} \\ 0 & \text{else} \end{cases} \tag{29}$$

where Q represents the pheromone strength, and its value affects the convergence speed of the algorithm. L_k represents the total cost of the path taken by the k-th ant in this cycle.

6. Case Analysis

In this paper, the IEEE-33 standard node distribution network is used as an original network for analysis [21]. The distribution network structure is shown in Figure 9.

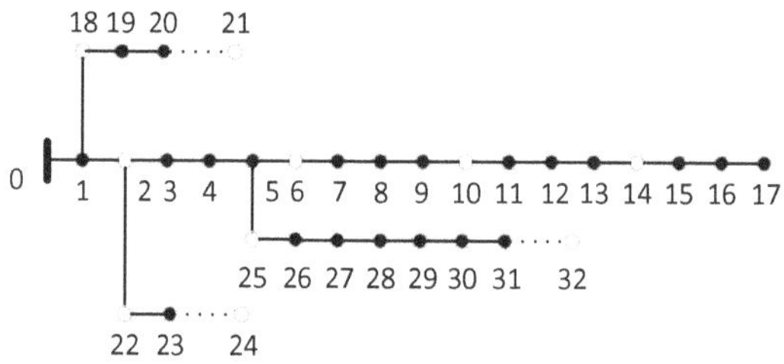

Figure 9. IEEE-33 standard node distribution network.

6.1. Case Parameters

In Figure 9, 0-33 is the node number, and L1-L33 are the line numbers. The node and line basic data are shown in Table 1.

Table 1. Node and line data.

Node Number	Load/MW	Load Type	Line Number	Start Node	End Node	R	X
0	0	balance					
1	0.1	commercial	1	0	1	9.22	4.7
2	0.09	commercial	2	1	2	4.93	25.1
3	0.12	commercial	3	2	3	36.6	18.6
4	0.06	commercial	4	3	4	38.1	19.4
5	0.06	resident	5	4	5	81.9	70.7
6	0.2	resident	6	5	6	18.7	61.8
7	0.2	resident	7	6	7	71.1	23.5
8	0.06	resident	8	7	8	103	74.1
9	0.6	resident	9	8	9	104	74.1
10	0.45	resident	10	9	10	19.6	6.5
11	0.06	resident	11	10	11	37.4	12.4
12	0.06	resident	12	11	12	146	115
13	0.12	resident	13	12	13	54.1	71.3
14	0.06	resident	14	13	14	59.1	52.6
15	0.06	resident	15	14	15	74.6	54.5
16	0.06	resident	16	15	16	128	172
17	0.09	commercial	17	16	17	73.2	57.4
18	0.09	commercial	18	1	18	16.4	15.6
19	0.09	commercial	19	18	19	150	136
20	0.09	commercial	20	19	20	40.9	47.8
21	0.09	commercial	21	20	21	0.819	-
22	0.09	commercial	22	2	22	45.1	30.8
23	042	commercial	23	22	23	89.8	70.9
24	0.42	commercial	24	23	24	0.374	-
25	0.06	resident	25	5	25	20.3	10.3
26	0.06	resident	26	25	26	28.4	14.5
27	0.06	resident	27	26	27	10.5	93.3
28	0.12	resident	28	27	28	80.4	70.1
29	0.2	resident	29	28	29	50.8	25.9
30	0.15	commercial	30	29	30	97.4	96.3
31	0.21	commercial	31	30	31	31.1	36.2
32	0.6	commercial	32	31	32	0.381	-

Based on the original AC distribution network, the following changes were made to simulate the AC/DC distribution network: the AC lines between node 20 and node 21, node 23 and node 24, node 31 and the node 32 change into the DC lines. Capacitors with a value of 1 are modified at nodes 20, 21, 23, 24, 31, 32. The VSCs are installed after the three DC lines, and the configuration parameters of the converter station are consistent: the rated capacity is 2MVA, the resistance is 0.5 Ω, and the reactance is 1.5 Ω.

The MT candidate nodes are 2, 6, 10, and 14, the PV candidate nodes are 21, 24, and 32, and the WG candidate nodes are 22 and 25. The maximum current carrying capacity of the AC line is 5000 kVA, and the maximum current carrying capacity of the DC line is 2500 kVA. The node voltage is allowed to range from 0.9 to 1.1 p.u., and the three types of DG are rated at 0.1 MW. The number of access nodes to be selected for all three DGs is limited to 15. The discount rate r is taken as 0.1, the DG recovery period n_y is taken as 20, the unit net loss c_{loss} is taken as 0.017 million yuan/(MW h), and the greenhouse gas emission intensity per unit MT K_{mt} is taken as 724.6 kg/(MW h). The environmental protection value of greenhouse gases V_{CO_2} is taken as 0.023 yuan/kg, and the fine for greenhouse gas emissions from MT R_{CO_2} is taken as 0.01 yuan/kg. The parameters of the genetic algorithm are: the population individual is 100, the number of iterations is 20, the crossover rate is 0.5, and the mutation rate is 0.1. The ant colony algorithm parameters are: the number of ants is 60, the pheromone concentration retention coefficient is 0.9, and the information heuristic factor and the expected heuristic factor are both 1, at the initial moment $\Delta\tau_{ij}^{k}(t) = 0$, the number of iterations is 60.

According to [22], it is assumed that DG's investment, operation and maintenance, fuel cost and environmental subsidies are as shown in Table 2.

Table 2. DG cost.

DG Type	Maintenance 10,000 Yuan/(MW·h)	Investment 10,000 Yuan/MW	Fuel 10,000 Yuan/(MW·h)	Subsidy 10,000 Yuan/(MW·h)
MT	0.01	48	0.06	0
PV	0.02	100	0	0.036
WG	0.03	63	0	0.01

6.2. Result Analysis

(1) Consider Timing Characteristics

By simulating the timing characteristics of DG and the load, this paper draws a plan that is more in line with the actual operation of the AC/DC distribution network. If the timing characteristics of DG and load are not considered, the output of DG in the system remains unchanged, and the output is based on the rated capacity. The two planning schemes considering the timing characteristics and disregarding the timing characteristics are shown in Tables 3 and 4.

Table 3. Planning schemes.

DG Type	Consider Timing Characteristics	Disregarding Timing Characteristics
MT	2(11), 6(15), 10(11), 14(4)	/
PV	21(2), 24(3), 32(8)	21(12), 24(14), 32(14)
WG	22(5), 25(1)	22(1), 25(7)

2(11) means that 11 DGs are installed at node 2, and so on.

Table 4. Costs of planning schemes.

DG Type	Consider Timing Characteristics	Disregarding Timing Characteristics
Maintenance cost/10,000 yuan	679.9	884.7
Investment cost/10,000 yuan	358.3	444.1
Loss cost/10,000 yuan	203.3	267
Fuel cost/10,000 yuan	1961	0
Pollution compensation cost/10,000 yuan	78.75	0
Subsidy cost/10,000 yuan	−417.6	−1322.8
Total cost/10,000 yuan	2863.7	273.1

It can be seen from the above results that the algorithm avoids including the MT if the timing characteristics are not considered. This is because although the investment and maintenance cost of gas MT are low, the environmental cost is too high, and the environmental subsidies for PV and WG are very high, the overall cost of MT is higher than that of PV and WG. However, the fact that the gas turbine is not added is obviously contrary to the actual situation, because in fact, the PV and WG outputs have strong volatility. If only these two types of DG are installed, there will be a certain period of time when the power supply does not meet the demand and this affects the reliable operation of the power grid. Therefore, it is necessary to consider the timing of DG.

(2) Algorithm Comparison

In order to verify the effectiveness of the GA-ACO algorithm, the genetic algorithm, ant colony algorithm and GA-ACO algorithm were used to optimize the DG access to the AC/DC distribution network. Since the results calculated by each of the algorithms in each iteration are different, this study performed 10 experiments on each of the three algorithms and took the average of the results. The final costs of the three algorithms are shown in Table 5 and the optimization curve for the three optimization algorithms is shown in Figure 10.

Table 5. Average cost of the three algorithms.

Planning	GA-ACO	GA	ACO
Maintenance cost/10,000 yuan	679.9	589.2	612.9
Investment cost/10,000 yuan	358.3	278.7	298.3
Loss cost/10,000 yuan	203.3	259.3	241.7
Fuel cost/10,000 yuan	1961	2231.6	2193.9
Pollution compensation cost/10,000 yuan	78.75	109.3	92.6
Subsidy cost/10,000 yuan	−417.6	−285.6	−301.8
Total cost/10,000 yuan	2863.7	3182.5	3137.6

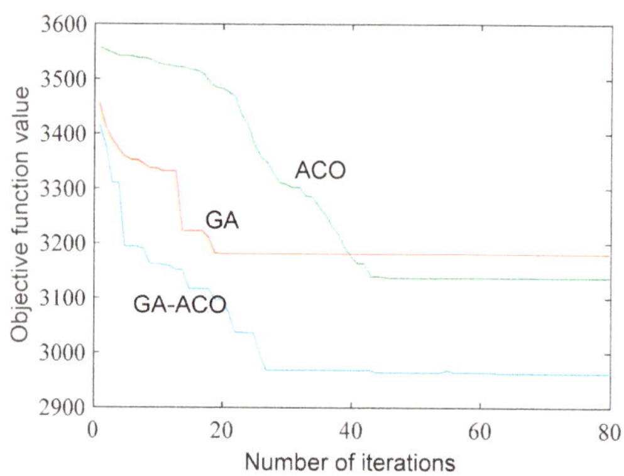

Figure 10. Optimization algorithm comparison chart.

From the table and the figure we can see that the genetic algorithm at the beginning of the optimization results is better than the ant colony algorithm, and the convergence speed is faster than the ant colony algorithm before the 20th generation, because the genetic algorithm has a wide range of search capabilities at the beginning of the search. The ant colony algorithm takes a long time to accumulate pheromones at the beginning. In the 20th generation of the iterations, the redundant iterations generated by the genetic algorithm in the later stage affect the convergence speed and fall into the local optimal solution. The convergence speed of the ant colony algorithm starts to accelerate, and the convergence is completed at about 40 generations. However, because the initial pheromone accumulation is not complete, the global optimal solution cannot be searched. The GA-ACO algorithm combines the advantages of both algorithms. At the beginning, it uses a wide range of genetic algorithms to search for pheromones, and then uses the ant colony algorithm to help the whole optimization process maintain a faster convergence speed and not fall into local parts optimal, then, the global optimal solution is found around the 60th generation.

(3) Considering Load Growth

Load growth is also taken into consideration in this paper. It is assumed that all loads will increase by 50% in the next 15 years [23]. The number of access nodes to be selected for all the three DGs will increase to 30. What is more, the line capacity will double. The simulation results are shown in Table 6 and Figure 11.

Table 6. Planning schemes considering load growth.

DG Type	Consider Timing Characteristics	Disregarding Timing Characteristics
MT	2(12), 6(27), 10(11), 14(7)	/
PV	21(8), 24(10), 32(15)	21(13), 24(18), 32(20)
WG	22(18), 25(8)	22(5), 25(3)

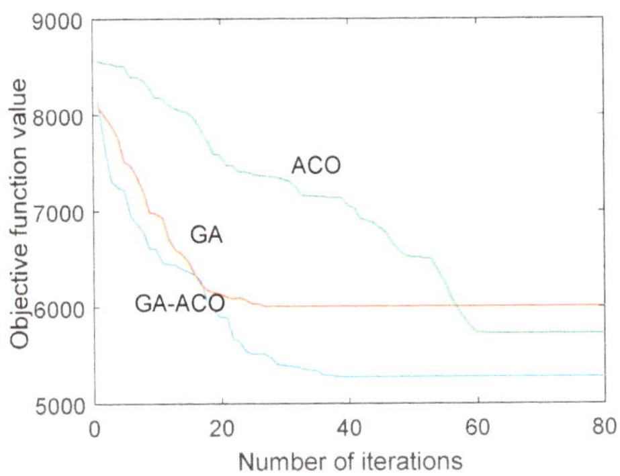

Figure 11. Optimization algorithm comparison chart considering load growth.

It can be seen that by considering the load growth, the model can still reasonably reflect the actual situation and this further verifies the effectiveness of the hybrid algorithm in convergence speed and convergence.

7. Conclusions

This paper aims to optimize the location and volume of DG in the AC/DC distribution network. Based on the consideration of DG and load timing characteristics, a DG planning model is established. The model is solved by the GA-ACO algorithm. The following conclusions were obtained from the case analysis:

(1) It is necessary to consider environmental costs in the problem of DG access to the distribution network, because this reflects the environmental advantages of PV and WG, helps to achieve a balance between system economics and environmental protection, and greatly improves the utilization rate of new energy, and this makes the distribution network cleaner and more environmental-friendly.
(2) Timing characteristics models are critical in solving the problem of DG access to the distribution network. By considering the timing characteristics, the model can more accurately reflect the operation of the distribution network, and the resulting planning scheme is more realistic.
(3) The GA-ACO algorithm combines the advantages of genetic and ant colony algorithms. It is superior to the two algorithms in terms of convergence speed and convergence results, and effectively avoids the objective function falling into the local optimal solution.

Author Contributions: Conceptualization, D.Y.; Formal analysis, F.M.; Writing—Original Draft, D.Y.; Writing—Review & Editing, F.M. and J.Z.

Funding: This paper is supported by the 2107 Key Research and Development Plan of Jiangsu Province BE2017030 "Research on Custom Power System in Photovoltaic/Battery microgrid with VSG interface".

Conflicts of Interest: The authors declare no conflict of interest.

References

1. Georgilakis, P.S.; Hatziargyriou, N.D. Optimal Distributed Generation Placement in Power Distribution Networks: Models, Methods, and Future Research. *IEEE Trans. Power Syst.* **2013**, *283*, 3420–3428. [CrossRef]
2. Boroyevich, D.; Cvetković, I.; Dong, D.; Burgos, R.; Wang, F.; Lee, F. Future electronic power distribution systems—A contemplative view. In Proceedings of the 2010 12th International Conference on Optimization of Electrical and Electronic Equipment, Basov, Romania, 20–22 May 2010; pp. 1369–1380.
3. Bifaretti, S.; Zanchetta, P.; Watson, A.; Tarisciotti, L.; Clare, J.C. Advanced Power Electronic Conversion and Control System for Universal and Flexible Power Management. *IEEE Trans. Smart Grid* **2011**, *22*, 231–243. [CrossRef]
4. Jiang, D.; Zheng, H. Current Status and Prospects of DC Distribution Network Research. *Autom. Electr. Power Syst.* **2012**, *368*, 98–104.
5. Zhao, B.; Zhao, Y.; Wang, Y.; Liu, G.W.; Song, Q. Energy Internet System Based on Flexible Medium Voltage DC Power Distribution. *Proc. CSEE* **2015**, *3519*, 4843–4851.
6. Elkhattam, W.; Hegazy, Y.G.; Salama, M.M.A. An Integrated Distributed Generation Optimization Model for Distribution System Planning. *IEEE Trans. Power Syst.* **2005**, *202*, 1158–1165. [CrossRef]
7. Zeng, B.; Liu, N.; Zhang, Y.; Yang, X.; Zhang, J.; Liu, W. Bi-level Scenario programming of active distribution etwork for promoting intermittent distributed generation utilization. *Trans. China Electrotech. Soc.* **2013**, *289*, 155–163.
8. Li, H.; Yang, J.; Cui, J. Research on the most preferable location and constant volume of multi-distributed power supply considering network loss and voltage. *J. Electr. Eng.* **2015**, *109*, 79–86.
9. Zhu, D.; Broadwater, R.P.; Tam, K.S.; Seguin, R.; Asgeirsson, H. Impact of DG placement on reliability and efficiency with time-varying loads. *IEEE Trans. Power Syst.* **2006**, *211*, 419–427. [CrossRef]
10. Guo, J.; Li, X.; Deng, W.; He, C.; Liu, W. Comprehensive optimal allocation of intermittent distributed generation and reactive power compensation based on bilevel planning. *Proc. CSEE* **2013**, *3328*, 25–33.
11. Rau, N.S.; Wan, Y.H. Optimum location of resources in distributed planning. *IEEE Trans. Power Syst.* **2002**, *94*, 2014–2020. [CrossRef]
12. Wang, C.S.; Chen, K.; Xie, Y.H.; Zheng, H.F. Siting and sizing of distributed generation in distribution network expansion planning. *Autom. Electr. Power Syst.* **2006**, *303*, 38–43.
13. Mei, H.; Cheng, H.; Wang, W. Optimal Configuration of Distributed Power Supply in Active Distribution Network. *J. Electr. Power Syst. Autom.* **2016**, *2811*, 1–8.
14. Ye, D.; He, Z.; Pei, T. Siting and sizing of distributed generation planning based on adaptive mutation particle swarm optimization algorithm. *Power Syst. Technol.* **2011**, *356*, 155–160.
15. Li, Y.; Mu, Q.; An, N. Development and Challenges of DC Grid Model and Simulation. *Autom. Electr. Power Syst.* **2014**, *384*, 127–135.
16. Frank, S.M.; Rebennack, S. Optimal design of mixed AC–DC distribution systems for commercial buildings: A Nonconvex Generalized Benders Decomposition approach. *Eur. J. Oper. Res.* **2015**, *2423*, 710–729. [CrossRef]
17. Wang, S.; Chen, S.; Xie, Y. Coordinated Economic Dispatch of Energy Storage and Converter Stations in AC/DC Distribution Network Considering Safety Constraints. *Autom. Electr. Power Syst.* **2017**, *4111*, 85–91.
18. Wang, Y.; Han, X.; Sun, D.; Li, Y.; Chen, Y.; Li, D. Power flow algorithm based on AC/DC associated minimum Jacobian matrix structure. *Autom. Electr. Power Syst.* **2015**, *39*, 1–6.
19. Xiao, H.; Pei, Y.; Deng, W. Analysis of the Influence of Distributed Power Supply on Distribution Network Voltage and Its Optimal Control Strategy. *Trans. China Electrotech. Soc.* **2016**, *31*, 203–213.
20. He, N. Comparative Analysis of Diurnal Variations of Wind Speed and Light Intensity in the Bohai Bay. *Meteorology* **1980**, *64*, 30–31.

21. Baran, M.E.; Wu, F.F. Network reconfiguration in distribution systems for loss reduction and load balancing. *IEEE Trans. Power Deliv.* **1989**, *42*, 1401–1407. [CrossRef]
22. Guo, X.; Cheng, H.; Masoud, B. Optimal sitingand sizing of distributed generation in microgrid based on adaptive DPSO. *Water Resour. Power* **2013**, *9*, 220–224.
23. Shao, Y.; Zhao, J.; Fang, J. Joint Dynamic Programming of Distributed Wind Power and Capacitor Based on Load Classification Growth. *Power Syst. Technol.* **2019**, *1*, 316–324.

applied sciences

MDPI

Article

A Stability Preserving Criterion for the Management of DC Microgrids Supplied by a Floating Bus

Daniele Bosich [1,*], Andrea Vicenzutti [1], Samuele Grillo [2] and Giorgio Sulligoi [1]

1 Department of Engineering and Architecture, University of Trieste, 34127 Trieste, Italy; avicenzutti@units.it (A.V.); gsulligoi@units.it (G.S.)

2 Dipartimento di Elettronica, Informazione e Bioingegneria, Politecnico di Milano, 20133 Milano, Italy; samuele.grillo@polimi.it

* Correspondence: dbosich@units.it; Tel.: +39-040-558-7123

Received: 30 September 2018; Accepted: 28 October 2018; Published: 1 November 2018

check for updates

Abstract: Direct current (DC) distribution is one of the most important enabling technologies for the future development of microgrids, due to the ease of interfacing DC components (e.g., batteries, photovoltaic systems, and native DC loads) to the grid. In these power systems, the large use of controlled power converters suggests the need of a careful analysis of system stability, as it can be impaired in particular conditions. Indeed, in DC power systems, a destabilizing effect can arise due to the presence of inductor/capacitor (LC) filtering stages (installed for power quality requirements) and high-bandwidth controlled converters, behaving as constant power loads (CPLs). This issue is even more critical when the CPL is potentially fed only by the battery, causing the DC bus to be floating. In this context, Lyapunov theory constitutes a valuable method for studying the system stability of DC microgrids feeding CPLs. Such a theory demonstrates how the region of asymptotic stability (RAS) shrinks as the state of charge of the battery diminishes (i.e., as the bus voltage decreases). Once the accuracy of the RAS is validated by comparing it to the real basin of attraction (BA), numerically derived using continuation methods, a smart power management of the CPL can be proposed to preserve the system stability even in the presence of a low bus voltage. Indeed, a suitably designed criterion for limiting the load power can guarantee the invariance of RAS and BA for each equilibrium point. An electric vehicle was used herein as a particular DC microgrid for evaluating the performance derating given by the power limitation.

Keywords: islanded DC microgrid; battery-only operation; constant power load; CPL; floating DC bus; stability criterion; power management

1. Introduction

Currently, direct current (DC) distribution represents the most innovative solution for islanded microgrids, both in transportation [1,2] and in land-based power systems [3,4]. Indeed, thanks to an even more widespread employment of power electronics and performing control systems, DC technology is not only capable of guaranteeing paramount advantages [5] (e.g., improved efficiency, enhanced power-flow control, and increased power availability), but it can also foster the optimal combination of generation, storage, and consumption [6]. Such a trend can be observed in the transportation industry [7,8] and in terrestrial systems [9,10], where many key factors and technologies are promoting this shift toward DC power systems, among which distributed energy resources (DERs) and energy storage systems (ESSs) are the most important. In fact, with ESSs and the great majority of DERs having a DC interface, the exploitation of this technology in some parts of the alternating current (AC) distribution system is becoming profitable. Moreover, several loads (e.g., data center, LED-based lighting systems, consumer electronics, etc.) are natively in DC, thus simplifying the transition. For characterizing the context in which this study is developed, a typical DC microgrid consists of a DC bus to which all components are connected through either DC–DC or DC–AC power converters. Such a DC bus is interfaced to the main AC grid through an AC–DC converter, whose main task is to control the bus voltage while balancing the power flow with the AC grid. The interactions between all these converters are usually investigated to identify possible interferences in the presence of controller bandwidth overlap.

In this context, a critical eventuality studied in academia is the constant power load (CPL) behavior [11,12], which may occur when a tightly controlled power converter (modeled as a nonlinear CPL) interacts with the inductor/capacitor (LC) filter, which is installed for ensuring the power quality requirements in DC systems. Depending on LC values and CPL power, the resonance among the nonlinear load and filtering components can jeopardize the DC microgrid voltage stability. During the past few years, several control techniques were proposed for solving such an instability [13–20], whereas the impedance-based stability criteria evaluated the impact of the nonlinear destabilizing CPL [21–23]. A different approach is based on Lyapunov theory [24–31], which overcomes the limits of small-signal linearization by determining the sufficient region of asymptotic stability (RAS) nearby a stable operating point. Conversely, other methods (e.g., numerical continuation analysis) can provide the actual basin of attraction (BA), as described in Reference [32], i.e., the sufficient and necessary area inside which the states can move without impairing the stability of the system. The two methodologies for obtaining RAS and BA were previously compared in Reference [33], demonstrating how the Lyapunov analytical formulation is more suitable for an online implementation aimed at guaranteeing the microgrid's stable operation.

Starting from the last conclusion, the present paper proposes a load management method for preserving system stability in critical DC power systems. The case under study is a particular islanded DC microgrid, where the floating bus is set only by the energy storage (without any active regulation) and the supplied load is a nonlinear CPL. For such a system, the direct dependency between the decrease of the battery state of charge (SoC) [34] and the shrinking of the RAS [33] suggests employing a CPL management for maintaining a proper stability margin.

The paper is organized as follows: Section 2 describes the DC microgrid topology, together with the related circuit in battery-only operation. In this section, the effect of the floating DC bus on system stability is analyzed by means of the Lyapunov theory, whose results (i.e., the RASs) are validated using a numerical continuation analysis (i.e., the BAs) and dynamic simulations. Section 3 proposes a stability-preserving criterion based on the definition of a stability index. In particular, this criterion is based on the mathematical law that describes how the CPL power depends on this index and on the DC bus voltage. This allows dynamically limiting the maximum available power in order to maintain the same stability margins, i.e., same RAS and BA, while the battery voltage decreases. The applicability of the proposed management is discussed in Section 4, where an electric vehicle is chosen as a DC microgrid example for testing the decrease in performance. Section 5 provides concluding remarks.

2. Effect of Floating DC Bus on System Stability

This section discusses the stability degradation that can possibly arise in a floating DC bus microgrid supplying a CPL. The present paper is based on a previous work [33] published by the authors about the criticality of a DC floating bus powering a hybrid electric vehicle. Nevertheless, the same conclusions can be extended to any islanded DC microgrid with a floating bus, supplying a high-bandwidth destabilizing load. In this regard, the simplified microgrid under study is described in Section 2.1, while initial considerations about the effect of a floating DC bus on the stability margin are proposed in Section 2.2. Then, two methodologies for large-signal analysis are introduced in Section 2.3, whereas numerical simulations are proposed in Section 2.4 for validating, through the system dynamic response analysis, the methodological approach used for the stability evaluation. In Section 2.5, the dynamic numerical simulations are used to verify the basin of attraction (BA) and demonstrate the validity of the Lyapunov analytical approach.

2.1. DC Microgrid Topology

The simplified power system shown in Figure 1 was chosen as a case study to analyze the voltage stability in a floating-bus DC microgrid in presence of a destabilizing CPL. The islanded microgrid used in this study is based on the following elements:

- DC power-generating system, composed of an internal combustion engine (IC), an alternator (G), and a controlled rectifier (P), to supply loads and/or recharge batteries;
- Energy storage system, i.e., an electrochemical battery (B);
- LC input filter (F), to assure proper voltage and current quality on the load bus;
- Generic static DC load (L), fed by a DC–DC converter (C);
- Generic rotating load (M), supplied by means of a controlled inverter (I).

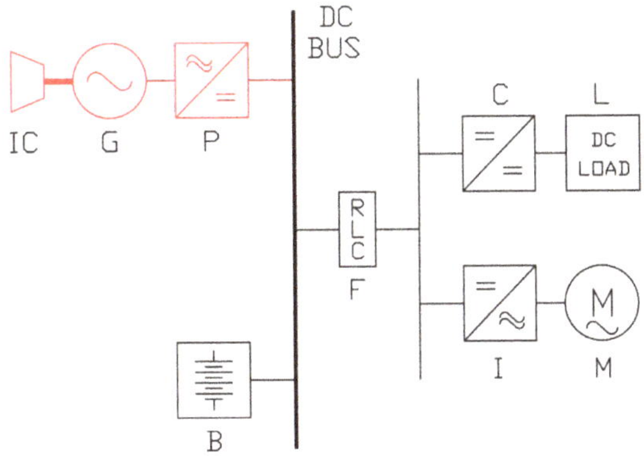

Figure 1. Proposed direct current (DC) microgrid; the power system section in battery-only operation is shown in black.

In order to save space, the system avoids using a battery charger unit, relying upon the controlled rectifier (P) to regulate battery charging. Thus, the DC bus voltage is floating when the power generating system is offline, depending on the battery SoC (the battery is the only source of power). As already shown in Reference [33], in this case voltage stability issues may affect the DC bus. Particular attention has to be paid to the loads; depending on the control bandwidth of the converters used to supply loads from the DC bus, it is possible to classify them into two classes: (i) conventional loads (CLs), and (ii) constant power loads (CPLs). The former category represents loads that do not require strict voltage regulation for proper operation (e.g., resistive heaters, loads with an integrated input conversion stage, etc.). Therefore, their global effect can be modeled through an equivalent linear resistance R_L. Conversely, the latter category is made up of loads requiring tight control by their input power converter (C or I in Figure 1), to either provide a constant voltage (for static loads) or a constant speed/torque (for rotating loads). The final effect is a constant power delivery from the DC bus to the loads. These loads can be modeled through a single equivalent nonlinear current $I_L = P/V$, where P is the CPL power. It has to be noticed that both static and rotating loads can be classified as CL or CPL, depending on their operating characteristics. The effect of several CLs and CPLs can be modeled using two equivalent aggregated loads. Given the interest in evaluating voltage stability in a floating-bus system, the simplified microgrid shown in Figure 1 was considered as being supplied by batteries only (power system section depicted in black in Figure 1). Thus, it is possible to model the overall power system in battery-only operation using the equivalent circuit of Figure 2, where E is the battery voltage, V is the voltage on the CPL, L and C are the filtering stage components, and R is the inductor physical resistance.

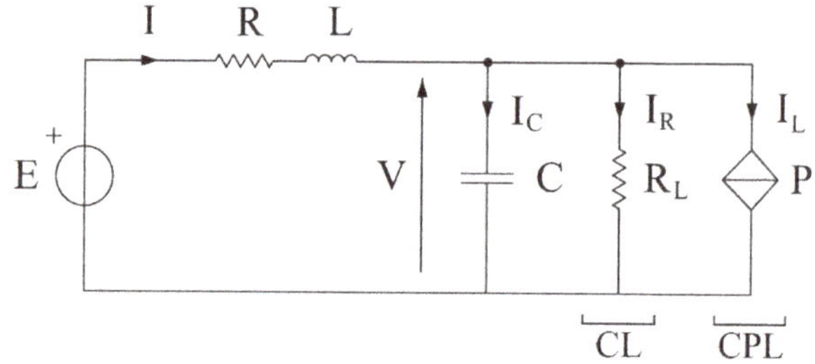

Figure 2. DC microgrid equivalent circuit model.

2.2. Definition of Hard Lower Bound for DC Load Voltage

The analysis in Reference [33] introduced the stability problem of a DC floating bus. In particular, an islanded power system supplied only by batteries may be affected by voltage instability in the presence of a perturbation capable of moving the system state outside the BA. This issue becomes critical when the battery has a low SoC and, consequently, its voltage output is low. In this case, the BA shrinks as the bus voltage decreases. This limits the operating margin of the power system.

To introduce the concept of hard lower bound for voltage stability, in this section, the case study was based on the data reported in Reference [33]. The adoption of per unit (p.u.) notation [34] allows performing a numerical continuation analysis [32,33], following the hypothesis of neglecting the presence of the CL (i.e., $R_L = \infty$). Indeed, the equivalent linear resistance R_L determines an increase in the damping factor, thus enhancing the system's voltage stability. Conversely, by neglecting CL, it is possible to assess the stability degradation in the worst case [35], thereby making the negative effect of the floating bus more apparent. The method used here to assess voltage stability relies on the evaluation of the basins of attraction (BAs) in the regions close to the stable operating point (Figure 3). Each stable operating point is defined by the couple of variables (v_0, i_0), where v_0 is the steady-state voltage on the capacitor (in p.u.), and i_0 is the steady-state nonlinear current in the filter inductor (in p.u.). The progressive reduction in the battery voltage e (due to the battery SoC decrease) causes the shift of the equilibrium point toward the upper left-hand side of Figure 3 (i.e., lower v_0 and higher current $i_0 = p/v_0$), with a consequent shrinking in the BA. In particularly, it can be noted that the progressive reduction in BA area (which can be assumed as a measure of system stability) becomes significantly faster for v_0 below 0.8 p.u. (refer to the area with a red boundary in Figure 3). Based on this consideration, it is possible to consider $v_0 = 0.8$ p.u. as a sort of hard lower bound for the steady-state voltage on the load bus (blue basin of Figure 3). Clearly, this limit on v_0 corresponds to a lower bound also for the battery voltage e, whose value depends on the voltage drop in the filter resistance component. Consequently, for each set of input data (i.e., filter components, CPL power, and rated voltage of the CPL), it is also possible to define a lower limit for the battery SoC, using the voltage limits and the battery specifications.

2.3. Basin of Attraction versus Region of Asymptotic Stability

The numerical continuation analysis is a method for studying the stability of nonlinear dynamic systems [32,33]. On the one hand, it allows identifying the Hopf bifurcation [36], which appears when the unstable limit cycle and the equilibrium point coalesce, thus making the equilibrium point unstable. Such a bifurcation was displayed at $v_0 = 0.696$ p.u. for the case studied in Reference [33]. On the other hand, the numerical continuation analysis makes it possible to evaluate the BA for several different equilibrium points. In fact, it allows determining the lower voltage stability bound once the magnitude of the possible perturbations is known. In particular, in Reference [33], the BA for $v_0 = 0.8$ p.u. was recognized as the smallest acceptable region for assuring stability in the presence of an impulse perturbation with a reasonable magnitude [37]. Actually, the defined lower bound allows keeping voltage stability after a perturbation constituted by an instantaneous voltage drop of up to ~30% with respect to the actual working point v_0 = 0.8 and i_0 = 1.25 p.u. (i.e., the system state moved to v = 0.5768 p.u. and i = 1.25 p.u.). This can be demonstrated by considering Figure 3, where the perturbed state (yellow triangle) is still in the calculated BA for the starting equilibrium point (blue-bounded area). Conversely, if a lower equilibrium point is assumed (e.g., v_0 = 0.7 p.u., resulting in the red-bounded area in Figure 3), the related BA is so small that the system can be considered unstable for any realistic perturbation.

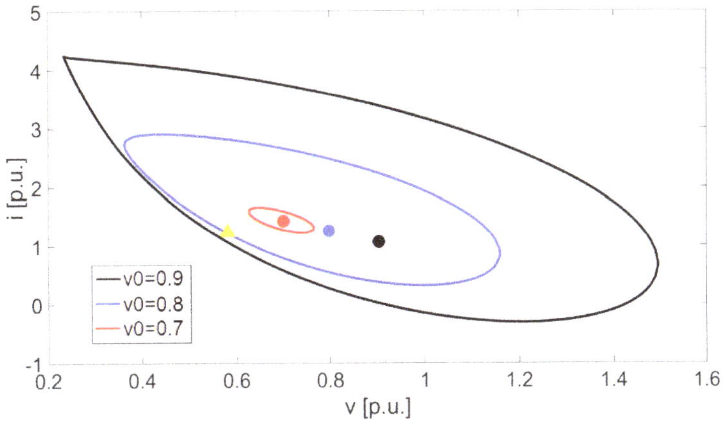

Figure 3. Basins of attraction (BAs) corresponding to shifted equilibrium points.

A different approach to define the stability limits of the system is to establish the region of asymptotic stability (RAS). By applying the Lyapunov theory [38], it is possible to find a conservative region, i.e., the RAS, where a sufficient but not necessary condition for large-signal stability is verified [30]. This, in turn, means that any system state inside the RAS originates a transient that evolves toward a stable equilibrium point. Conversely, system states located outside the RAS are not guaranteed to originate an unstable evolution. Although RAS provides only a subset of the entire BA, its determination is significant. Unlike the BA, which is found numerically, the RAS is defined analytically. Therefore, it is possible to use it in algorithms and control laws able to perform a stability-oriented real-time load management in floating-bus islanded DC microgrids. In particular, the mathematical approach at the basis of the definition of the RAS is used in this paper to define a stability preserving criterion, able to suitably reduce the CPL power when the bus voltage drops. In such a way, the voltage stability is guaranteed even in the presence of low battery voltage output e, i.e., low SoC, assuring the correct supply of the remaining loads.

2.4. Numerical Simulation

By removing the conventional load CL from the equivalent circuit in Figure 2, it is possible to define the system's states as in Equation (1).

$$\begin{cases} \dot{i} = \frac{di}{dt} = \frac{1}{l}(e - ri - v) \\ \dot{v} = \frac{dv}{dt} = \frac{1}{c}(i - i_L) = \frac{1}{c}\left(i - \frac{p}{v}\right) \end{cases}, \tag{1}$$

where v is the CPL voltage, and i is the current flowing in the inductor l. The system parameters were those used in Reference [33], while the CPL power was set at its rated value (i.e., p = 1 p.u.).

Using Matlab Simulink, Equation (1) was implemented to perform simulations to assess the v–i transients following a perturbation. The results were used to provide a dynamic validation of the stability limits previously calculated. The studied DC microgrid in steady-state condition (working point v_0 = 0.8 and i_0 = 1.25 p.u.) was perturbed at $t = t_0$ = 0.2 s by a voltage impulse capable of instantaneously moving the voltage v applied to the CPL to the new $v(t_0)$ voltage. After the perturbation, the state variables were free to evolve. The study developed in Reference [37] previously demonstrated that this perturbation can be employed for effectively testing the large-signal stability. Therefore, in the following sections, a voltage impulse was considered proper for evaluating the capability offered by the two approaches. The variation in the perturbed initial state $v(t_0)$ allows comparing the consequent v–i transients shown in Figures 4 and 5. In particular, it is possible to notice an unstable behavior when $v(t_0) = 0.56$ p.u. (black curves), whilst red/green transients ($v(t_0) = 0.59$ p.u.) are stable and converge toward the pre-disturbance working point. Thus, the performed simulations verified the lower voltage stability limit for the system calculated using BA analyses ($v \approx 0.57$ p.u.).

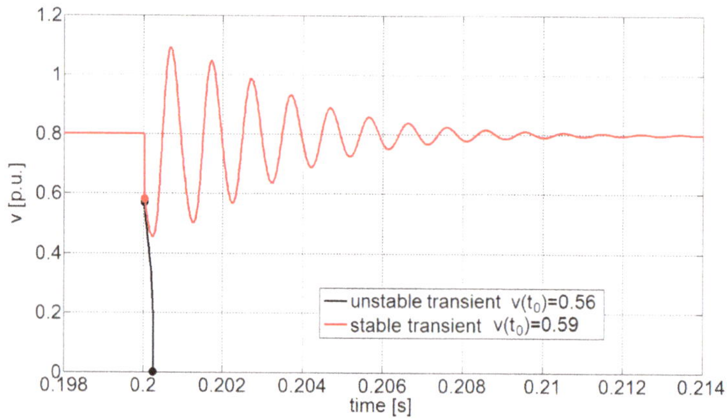

Figure 4. Voltage transients due to a voltage impulse perturbation at $t = 0.2$ s.

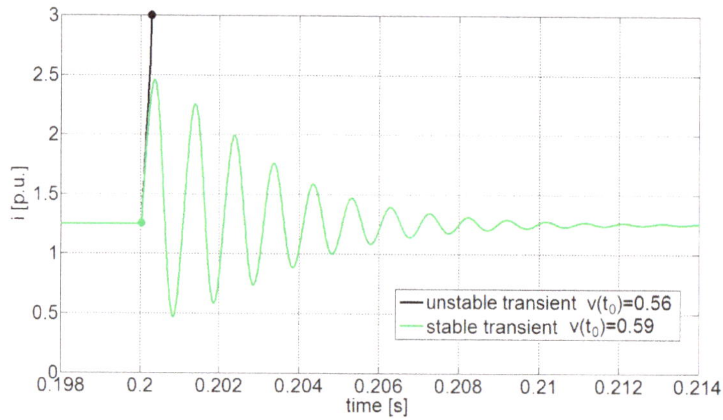

Figure 5. Current transients due to a voltage impulse perturbation at $t = 0.2$ s.

2.5. Validation of Methodological Approach

Dynamic transients are useful for validating both approaches with regards to the large-signal stability, i.e., BA and RAS. To this aim, BA and RAS for the hard lower bound case (i.e., working point $v_0 = 0.8$ p.u. and $i_0 = 1.25$ p.u.), together with the dynamic transients after the perturbation, are depicted in the v–i state plane of Figure 6. As can easily be seen, the BA (blue line) can correctly assess system stability. As expected, the transient starting from outside the basin at $v(t_0) = 0.56$ p.u. (black point) diverges, whereas the red trace of the transient starting inside the basin at $v(t_0) = 0.59$ p.u. (red point) converges toward the stable working point. Conversely, the RAS (green line) covers only part of the actual BA. This is expected, as the RAS is based only on a sufficient condition. Therefore, it is impossible to predict the system stability through Lyapunov analysis for the transient starting from $v(t_0) = 0.59$ p.u. (red point).

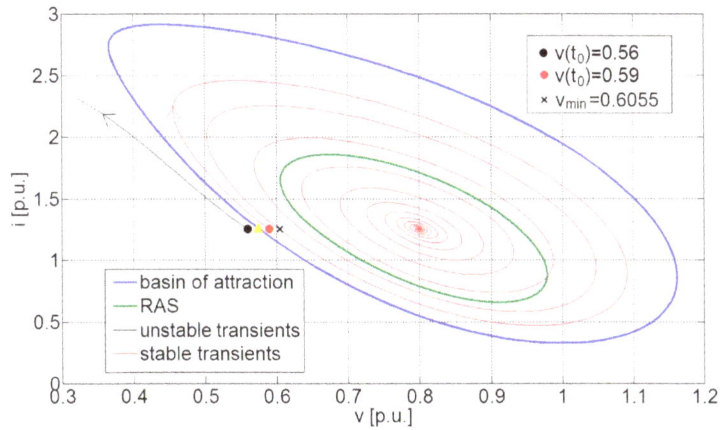

Figure 6. BA and region of asymptotic stability (RAS); validation when $v_0 = 0.8$ p.u.

However, although the RAS can only partially analyze the large-signal behavior, it can constitute a valuable method for designing a stability preserving criterion for floating-bus DC microgrids. Indeed, the determination of the RAS is possible after the identification of the voltage limit above which the Lyapunov first derivative (Equation (2)) is negative.

$$\dot{\Psi}(v) = \frac{1}{lc}\left\{v + rp\frac{1}{v} - e\right\} \cdot \left\{-\frac{r}{l}(v - v_0) - \frac{p}{c}\frac{1}{v} + \frac{p}{cv_0}\right\}. \tag{2}$$

For the microgrid under study, such a limit is expressed by Equation (3).

$$\dot{\Psi}(v) \leq 0 \Leftrightarrow v \geq \frac{pl}{rcv_0} := v_{\min}. \tag{3}$$

By assuming the system parameters used in Reference [33]—CPL power $p = 1$ p.u. and working point $v_0 = 0.8$ p.u.—the voltage limit $v_{\min} = 0.6055$ p.u. can be determined (represented by the "×" in Figure 6). Actually, $v_{\min}$ and the basin's voltage limit (yellow triangle located at $v = 0.5768$ p.u.) are very close, having a difference smaller than 0.03 p.u. Considering this gap negligible, the voltage limit $v_{\min}$ can act as an effective, yet still conservative, margin for the large-signal stability in the presence of the class of perturbations envisaged in Section 2.4. Moreover, it has to be noted that Equation (3) is a simple equation that can be evaluated immediately, thus making the voltage limit assessment very easy.

3. Stability Preserving Criterion

As previously observed, the system stability is negatively affected by the DC bus voltage decreasing in islanded microgrids feeding a CPL. A possibility for avoiding the consequent instability is to reduce the CPL. In this perspective, a suitably designed CPL management system may be useful for re-establishing a proper stability margin when the bus voltage (which depends on battery SoC in battery-only operation) is low. In order to define such a stability preserving criterion, it is firstly important to understand how the chosen stability index depends on both the CPL power and the floating-bus voltage (Section 3.1). Then, a strategy for decreasing the CPL power can be introduced (see Section 3.2), to guarantee a stable operation even when battery SoC is low. Finally, it is possible to assess the stability performance ensured by the proposed criterion (see Section 3.3) by verifying the presence of a sufficiently large stability region (either BA or RAS) in the v–i state plane as the SoC decreases, and the operating point moves consequently.

3.1. Stability Index

The Lyapunov analysis demonstrated the importance of the v_{min} term (Equation (3)) in a DC microgrid supplying a CPL. As shown in Figure 6, this parameter represents the lowest voltage margin for a specific DC power system (with given r, l, and c parameters) supplying a CPL with power p and working in steady state at the voltage v_0. In this regard, it is possible to define the distance Δ between the equilibrium point v_0 and the lower bound v_{min} as a conservative stability index. In fact, any perturbation capable of moving the voltage state inside the area defined by Δ does not jeopardize the system stability, as the Lyapunov conservative condition is still verified.

As previously mentioned, the battery voltage reduction (due to SoC decrease) is responsible for an operating point shift toward the upper left-hand side of the v–i state plane (v_0 decreases and i_0 increases), due to the relationship between supply voltage and absorbed current in a CPL. Moreover, the v_0 drop determines an increase in the v_{min} limit, as highlighted by Equation (3), which, in turn, leads to a further reduction of Δ. As this behavior is particularly important for the stability issue, the following mathematical study aims to demonstrate the relationship between the bus voltage decrease and the stability index Δ shrinking. To study this issue, it is necessary to define the parameter e_{0t}, which is the battery voltage needed to supply the CPL rated power (p = 1 p.u.) at the rated load voltage (v_0 = 1 p.u.). This parameter is representative of an optimistic scenario, with a fully charged battery. Conversely, in normal operating conditions, the battery SoC is lower, thus leading to a lower battery voltage e_0, which can be represented as a percentage ($e_{\%}$) of the full charge voltage e_{0t}. By observing Figure 2 and assuming the steady-state condition, the battery voltage e_0 is defined through Equation (4).

$$e_0 = v_0 + r\frac{p}{v_0} = e_{0t}e_{\%}. \tag{4}$$

By rearranging the second equality of Equation (4), it is possible to derive Equation (5), and then obtain the second-order Equation (6) by multiplying Equation (5) with the unknown quantity v_0.

$$v_0 - e_{0t}e_{\%} + r\frac{p}{v_0} = 0. \tag{5}$$

$$v_0^2 - e_{0t}e_{\%}v_0 + rp = 0. \tag{6}$$

By neglecting negative roots, the steady-state voltage v_0 results in Equation (7), while the total battery voltage e_{0t} is defined in Equation (8), once the rated condition ($p = v_0$ = 1 p.u.) and full battery ($e_{\%}$ = 1.0) are applied to Equation (4):

$$v_0 = \frac{e_{0t}e_{\%} + \sqrt{(e_{0t}e_{\%})^2 - 4rp}}{2}. \tag{7}$$

$$e_{0t} = 1 + r. \tag{8}$$

Assuming a variable $e_{\%}$, Equations (7) and (8) can be used to delineate the v_0 voltage shift in the presence of different load powers. Consequently, the lower bound v_{min} corresponding to each v_0 value can be determined using Equation (3), whereas the stability index Δ can be found with Equation (9).

$$\Delta = v_0 - v_{min} = v_0 - \frac{pl}{rcv_0}. \tag{9}$$

The equations explained so far allow the drawing of Figure 7, where the gradual contraction of Δ is made evident in response to $e_{\%}$ decrease. In particular, for the system under study, the v_0 curve intersects the $v_{\min}$ trace (i.e., $\Delta = 0$) for $e_{\%} \approx 0.83$ in the case of rated power (blue curve). At this specific point, both the large- and the small-signal stability are impaired.

$$\Delta = 0 \rightarrow v_0 = v_{\min} \rightarrow v_0 = \frac{pl}{rcv_0} \rightarrow v_0 = \sqrt{\frac{pl}{rc}}. \tag{10}$$

On the contrary, for smaller values of CPL power p (black/red curves), larger values of Δ are assured along the $e_{\%}$ drop, thus revealing a possible strategy for ensuring a sufficient stability margin. Actually, it is possible to conceive a smart management of the CPL, able to conveniently decrease its power p as the voltage e_0 decreases, in order to guarantee a proper stability margin.

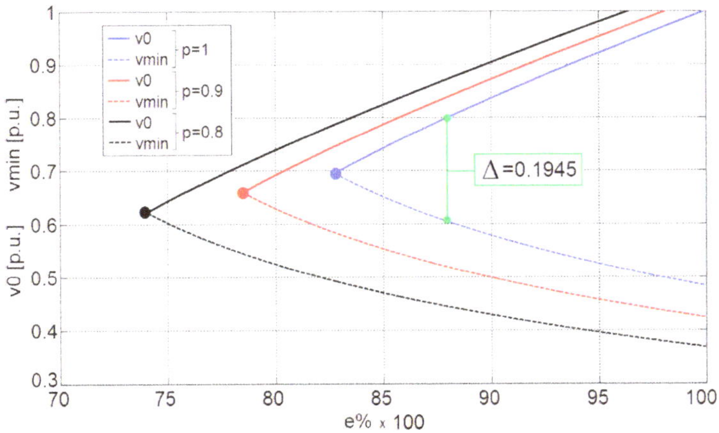

Figure 7. Stability index Δ, with varying constant power load (CPL) power and DC bus voltage.

3.2. Smart CPL Management

As discussed above, when the islanded microgrid is in battery-only operation, a smart management of the CPL power can be useful to preserve voltage stability. To this aim, this subsection focuses on the definition of a power reduction function able to ensure a proper stability index Δ while the battery is discharging (i.e., $e_{\%}$ is decreasing). By multiplying Equation (9) with the steady-state voltage v_0, a second-order equation can be obtained:

$$v_0^2 - \Delta v_0 - \frac{pl}{rc} = 0. \tag{11}$$

Discarding the negative root, the solution for (11) is

$$v_0 = \frac{\Delta}{2} + \sqrt{\frac{\Delta^2}{4} + \frac{pl}{rc}}. \tag{12}$$

The relationship between e_0 and Δ is

$$e_0 = v_0 + rp \cdot \left(\frac{\Delta}{2} + \sqrt{\frac{\Delta^2}{4} + \frac{pl}{rc}} \right)^{-1}. \tag{13}$$

By dividing/multiplying the second term of Equation (13) by the term

$$\left(\frac{\Delta}{2} - \sqrt{\frac{\Delta^2}{4} + \frac{pl}{rc}} \right), \tag{14}$$

a new equation is found as a special product (i.e., difference of two squares):

$$e_0 = v_0 + \frac{r^2 cp}{-pl}\left(\frac{\Delta}{2} - \sqrt{\frac{\Delta^2}{4} + \frac{pl}{rc}}\right). \tag{15}$$

Once the v_0 expression (Equation (12)) is substituted into Equation (15), one obtains

$$e_0 = \frac{\Delta}{2} + \sqrt{\frac{\Delta^2}{4} + \frac{pl}{rc}} - \frac{r^2 c\Delta}{2l} + \frac{r^2 c}{l}\sqrt{\frac{\Delta^2}{4} + \frac{pl}{rc}}, \tag{16}$$

$$e_0 + \left(\frac{r^2 c}{l} - 1\right) \cdot \frac{\Delta}{2} = \left(\frac{r^2 c}{l} + 1\right)\sqrt{\frac{\Delta^2}{4} + \frac{pl}{rc}}. \tag{17}$$

Then, by squaring Equation (17), Equation (18) is obtained and, consequently, after straightforward manipulation, Equations (19) and (20).

$$e_0^2 + \left(\frac{r^2 c}{l} - 1\right) \cdot \Delta e_0 + \left(\frac{r^2 c}{l} - 1\right)^2 \cdot \frac{\Delta^2}{4} = \left(\frac{r^2 c}{l} + 1\right)^2 \left(\frac{\Delta^2}{4} + \frac{pl}{rc}\right), \tag{18}$$

$$e_0^2 + \left(\frac{r^2 c}{l} - 1\right) \cdot \Delta e_0 + \left[\left(\frac{r^2 c}{l} - 1\right)^2 - \left(\frac{r^2 c}{l} + 1\right)^2\right] \frac{\Delta^2}{4} = \left(\frac{r^2 c}{l} + 1\right)^2 \frac{pl}{rc}, \tag{19}$$

$$e_0^2 + \left(\frac{r^2 c}{l} - 1\right)\Delta e_0 - \frac{r^2 c}{l}\Delta^2 = \left(\frac{r^2 c}{l} + 1\right)^2 \frac{pl}{rc}. \tag{20}$$

Finally, starting from

$$le_0^2 + \left(r^2 c - l\right)\Delta e_0 - r^2 c\Delta^2 = \left(r^2 c + l\right)^2 \frac{p}{rc}, \tag{21}$$

the function $p = f(e_0, \Delta)$ can be derived:

$$p = \frac{rc}{(r^2 c + l)^2}\left[le_0^2 + \left(r^2 c - l\right)\Delta e_0 - r^2 c\Delta^2\right]. \tag{22}$$

Equation (22) explains how to modify the CPL power p to maintain a desired Δ when e_0 becomes lower.

To better analyze Equation (22), the filter parameters r, l, and c may be gathered in the K term,

$$K = \frac{\tau_L}{\tau_C} = \frac{l}{r} \cdot \frac{1}{rc} = \frac{l}{r^2 c}, \tag{23}$$

through the definition of the time constants τ_L and τ_C. The K term allows simplifying Equation (22).

$$p = \frac{rc}{(r^2 c)^2 (1 + K)^2}\left[Kr^2 ce_0^2 + r^2 c(1 - K)\Delta e_0 - r^2 c\Delta^2\right]; \tag{24}$$

$$p = \frac{Ke_0^2 + (1 - K)\Delta e_0 - \Delta^2}{r(1 + K)^2}. \tag{25}$$

Finally, by substituting the e_0 voltage definition in Equation (4) into Equation (25), one obtains

$$p = \frac{e_{\%}^2\left[e_{0t}^2 K\right] + e_{\%}\left[\Delta e_{0t}(1 - K)\right] - \Delta^2}{r(1 + K)^2}. \tag{26}$$

Such a function, equivalent to Equation (22), better highlights the relationship between voltage and power. For this reason, it can be called the "power function". Through Equation (26), it is possible to appreciate the reduction in CPL power needed to migrate the working point (i.e., a decrease in i_0) toward a new point with the given (desired) stability index Δ (Figure 8). In particular, the green curve in Figure 8 is related to the power reduction function capable of guaranteeing the rated stability index, whereas greater/smaller values of Δ are ensured when the power reduction follows the red/blue curves. Clearly, if the requested stability index exceeds the rated value (0.6 vs. 0.516, red curve), then the rated power is not reachable even with a fully charged battery. Conversely, the DC microgrid can feed the rated CPL power with a partially charged battery (e.g., $e_{\%} = 0.95$) if the stability performance is downgraded (0.4 vs. 0.516, blue curve). As stated in References [11–20], the stability of a DC power system supplying a CPL is closely related to the filter parameters. For this reason, Figure 9 depicts the influence of the parameter K on the power function. In particular, by halving K (i.e., the capacitance c is doubled with respect to the inductance l, keeping the resistance r constant), the range of voltages in which the rated power can be supplied is extended (blue curve). On the contrary, more critical scenarios are given by a double K (red curve), where the DC microgrid can never provide the rated power to the CPL while keeping voltage stability (i.e., $p < 0.5$ p.u. when $e_{\%} = 0.9$).

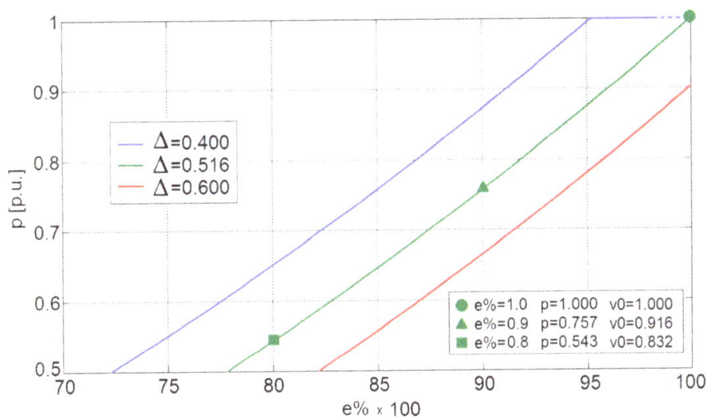

Figure 8. Power reduction function for preserving different values of stability index Δ.

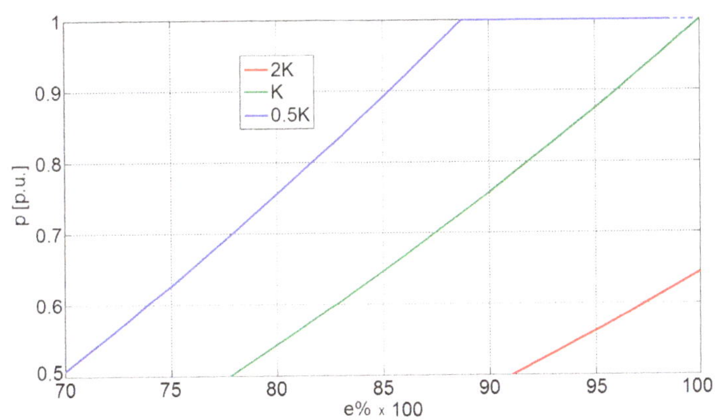

Figure 9. Power reduction function in the presence of different filtering solutions.

3.3. Migration of RAS and BA

The effects of the proposed power reduction function on RASs and BAs of the system are depicted in Figures 10 and 11, respectively. As expected, both RAS and BA shift along the v–i plane, as $e_{\%}$ and p diminish. Moreover, their area remains constant, due to the stability index invariance ($\Delta = 0.516$) guaranteed by applying the proposed power reduction function.

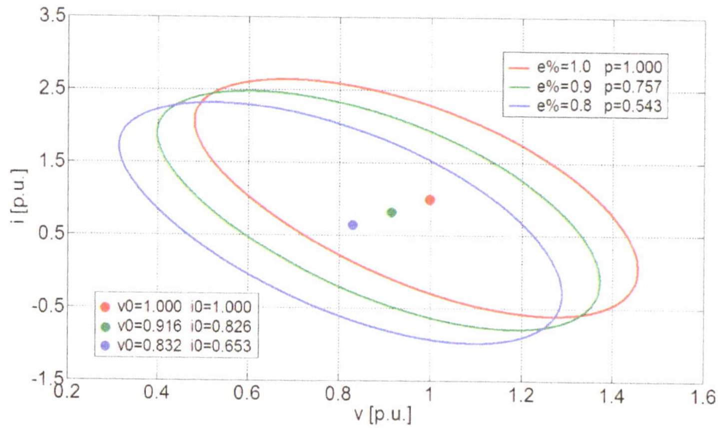

Figure 10. RAS migration imposed by power reduction function.

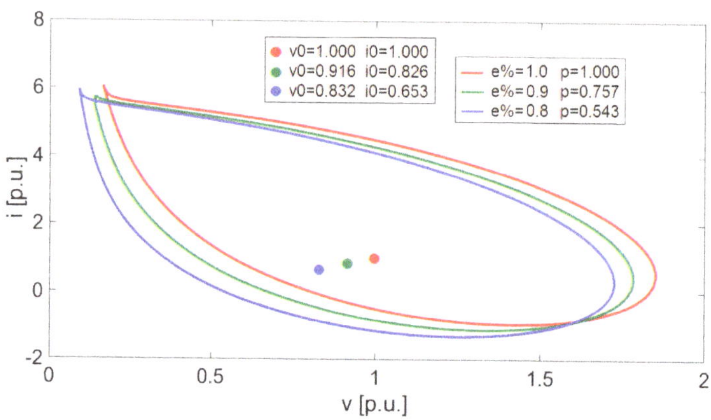

Figure 11. BA migration imposed by power reduction function.

4. Application Example

To demonstrate the applicability of the proposed smart CPL management (see Section 3.2) and to test the possible outcomes of the power derating on system performance, an application example was chosen. Specifically, a full electric vehicle was selected. The electric propulsion motor and its inverter constitute the CPL. Such a choice allows clearly showing the impact on performance of the proposed smart CPL management by evaluating indices that are evident and easy to understand (i.e., maximum achievable speed and 0–100 km/h acceleration time).

In order to study the worst case for DC voltage stability, the power system of the application example was simplified by removing the CLs and considering battery-only operation. Thus, a battery, an LC filter, and a CPL (the propulsion system) constituted the power system to be analyzed. The result was an islanded DC microgrid with a floating bus, similar to the one analyzed in the previous Sections.

The data of a commercial electric car (2011 Nissan Leaf) were inferred from several online datasheets, and are shown in Table 1. Using standard physics equations and applying a set of simplifying hypotheses, a Matlab script was developed in order to evaluate the performance of the car with respect to the CPL power. As previously mentioned, the maximum achievable speed and the 0–100 km/h acceleration time were chosen as performance indices. Proper modeling was applied to allow the calculation of these indices starting from available propulsion power. To lower the modeling burden, the vehicle was considered as a point of mass, free from the influence of external forces (e.g., no height variations and no wind), whereas the propulsion motor and inverter system had a control bandwidth so high with respect to the overall vehicle that time constants could be considered negligible. As previously discussed, CPL power has to be decreased through the application of the power reduction function (Equation (26)), to preserve the DC voltage stability when the battery discharge causes a decrease in the bus voltage. In particular, the maximum battery voltage range was defined by means of Table 1 parameters: 403 V voltage at full charge means $e_{\%} = 1$, while 336 V at 20% SoC corresponds to $e_{\%} = 0.83$. For what concerns the stability index, the value $\Delta = 0.516$ was chosen as a feasible tradeoff between a wide stability margin and the applicable CPL power (as shown in Figure 8, green trace). Using these data, the reduced CPL power can be determined through Equation (26), leading to the results depicted in Table 2. Focusing on the performance indices, the maximum achievable speed can be determined by exploiting the steady-state force equilibrium equation,

$$F_m = F_f + F_a, \tag{27}$$

where F_m is the force applied by the electric motor to the wheels, F_f the force due to the wheel–road friction, and F_a is the drag force due to the air. These three forces were assessed by considering the additional system parameters reported in Table 3. In particular, to model the overall transmission losses, the wheel traction was calculated by reducing the electric motor power through a 15% loss coefficient, thus resulting in a wheel power ranging from 37 to 68 kW (Table 2). Conversely, the wheel–road friction F_f and the air drag force F_a were determined using the following equations:

$$F_f = \mu_d RMg, \tag{28}$$

$$F_a = \frac{1}{2}\left(r_0 A C_x v^2\right), \tag{29}$$

where the parameters are defined in Tables 1–3, whilst v is the car speed in m/s, and g the gravity acceleration constant. Finally, the maximum achievable speed was calculated as a function of the power at the wheels using Equations (27)–(29), the results of which are shown in Figure 12 (red curve, left ordinate).

On the other hand, the 0–100 km/h acceleration time can be found by means of the dynamic equation:

$$a(v) = \frac{1}{2M}\left[F_{md}(v) - F_f - F_a(v)\right], \tag{30}$$

where $a(v)$ is the vehicle acceleration, $F_a(v)$ the drag force due to the air, and $F_{md}(v)$ is the dynamic force applied by the electric propulsion motor to the wheels.

In particular, by taking into account not only the propulsion motor's available power (Table 2), but also the maximum force transferrable from wheels to the asphalt in bad weather conditions and a traction control safety coefficient (refer to Table 3), the term F_{md} may be obtained.

$$F_{md}(v) = \begin{cases} F_m(v) & if\, F_m \leq F_{\lim} \\ F_{\lim} & if\, F_m > F_{\lim} \end{cases}, \tag{31}$$

$$F_{\lim} = \frac{2}{4} k_{tc} \mu_s g M, \tag{32}$$

where F_{lim} represents the maximum force transferrable from wheels to the asphalt by a four-by-two-wheel drive car. It is worth noting that, in Equation (30), all the terms depend on speed. Therefore, corresponding to a series of different propulsion power values, Equation (30) was used to calculate the vehicle speed variation from zero to its maximum. The resulting dataset, which relates vehicle speed and time for each power value, allows evaluating the 0–100 km/h time with a simple search algorithm. The results of this procedure are shown in Figure 12 (blue curve, right ordinate), where the acceleration time is shown with respect to the power available at the wheels.

By analyzing Figure 12, it is possible to evaluate the effect of the proposed smart CPL management on the performance indices for the application example. Firstly, the maximum achievable speed at the minimum available power is still in the range of the maximum speed limit set by the car software (135 km/h achievable vs. 144 km/h limit). Secondly, the acceleration performance at the lowest battery level is still satisfactory (~10.4 s) with respect to common car performance levels. In this regard, it has to be noted that these results were calculated in a worst-case condition, thus possibly leading to lower performance loss in a real system.

Although these results were obtained by oversimplifying both the power system and the system's physics (i.e., not constituting a complete performance assessment), they may provide important insight into the stability criterion effect. Actually, considering the application example, the proposed criterion can represent a valuable solution for preserving the DC stability by suitably managing the CPL power without excessively impairing the system performance. Obviously, the applicability of the proposed smart CPL management depends not only on the single system under study, but also on some parameters that can be defined accordingly by designers (such as coefficient K and stability margin Δ).

Table 1. Application example: electric vehicle parameters. SoC—state of charge.

Electric motor power	P_n	80 kW
Battery capacity		24 kWh
Nominal battery voltage		360 V
Battery voltage at full charge		403 V
Battery voltage at 20% SoC		336 V
Mass	M	1500 kg
Frontal area	A	2.28 m^2
Wheel diameter (205/55 R16)	R	63.16 cm
Maximum speed (unlimited)		170 km/h
Maximum speed (software limited)		144 km/h
0–100 km/h (unofficial tests)		~9 s
Aerodynamic penetration coefficient	C_x	0.32

Table 2. Application example: power reduction function effect and resulting power at the wheels.

Battery Voltage (%)	Electric Motor Power (per Unit)	Electric Motor Power (kW)	Wheels Power (kW)
100	1	80.00	68.00
95	0.875	70.00	59.50
90	0.757	60.56	51.48
85	0.646	51.68	43.93
80	0.543	43.44	36.92

Table 3. Application example: system parameters.

Static friction coefficient (rubber–asphalt) wet conditions	μ_s	0.5
Rolling friction coefficient (rubber–asphalt)	μ_d	0.035
Traction control safety coefficient	k_{tc}	0.75
Power loss (from motor to wheels)		15%
Minimum battery SoC		20%
Air density (kg/m^3)	r_0	1.29

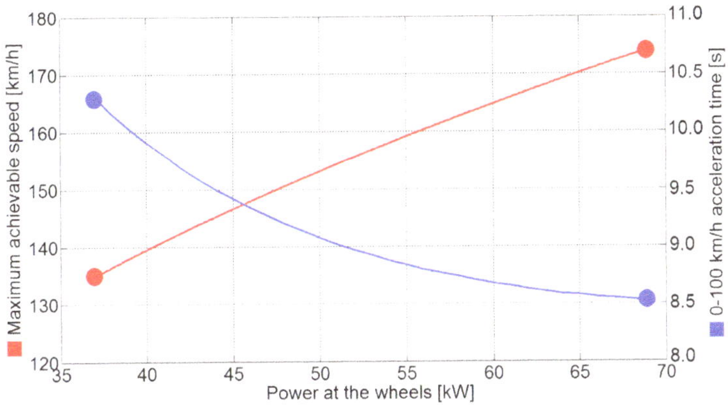

Figure 12. Maximum achievable speed and 0–100 km/h acceleration time as a function of power at the wheels.

5. Conclusions

The paper proposed a smart power limitation in order to preserve system stability in a DC microgrid feeding a CPL. Such a management is crucial when the battery is the only power source for the microgrid. The battery operation determines a floating DC bus, whose unregulated voltage varies over time depending on the SoC of the battery. When the battery SoC is decreasing, i.e., the DC bus voltage is diminishing, the stability region shrinking can be evaluated by means of the Lyapunov theory and numerical continuation analysis. The analytical method demonstrated its effectiveness in defining the stability index, which has the main role in the definition of the stability preserving criterion. For a given power-quality filter, the power function expresses the mathematical relationship among bus voltage, stability index, and CPL power. Therefore, a proper management can be designed to limit the power of the CPL to assure stability when the steady-state DC bus voltage decreases. This power reduction was analyzed on the basis of several results, which confirm the invariance of both the area and shape of the stability regions (i.e., RASs and BAs), while the operating point moves in the state plane. The paper finally demonstrated the effectiveness of the proposed criterion by means of a suitable application example.

Author Contributions: D.B. conceived the presented idea and developed the theory. S.G. verified the analytical method. A.V. developed the case study and G.S. supervised the research activity. All the authors discussed the results and contributed to the final manuscript.

Funding: This research received no external funding.

Acknowledgments: Authors wish to acknowledge Giovanni Giadrossi for the valuable contributions given in revising this work.

Conflicts of Interest: The authors declare no conflicts of interest.

References

1. Hansen, J.F.; Wendt, F. History and State of the Art in Commercial Electric Ship Propulsion, Integrated Power Systems, and Future Trends. *Proc. IEEE* **2015**, *103*, 2229–2242. [CrossRef]
2. Emadi, A. Transportation 2.0. *IEEE Power Energy Mag.* **2011**, *9*, 18–29. [CrossRef]
3. Zubieta, L.E. Are Microgrids the Future of Energy? DC Microgrids from Concept to Demonstration to Deployment. *IEEE Electrif. Mag.* **2016**, *4*, 37–44. [CrossRef]
4. Patterson, B.T. DC, Come Home: DC Microgrids and the Birth of the "Enernet". *IEEE Power Energy Mag.* **2012**, *10*, 60–69. [CrossRef]
5. IEEE Std. 1709-2010. *IEEE Recommended Practice for 1 to 35 kV Medium Voltage DC Power Systems on Ships*; IEEE: Piscataway, NJ, USA, 2010.

6. Meng, L.; Shafiee, Q.; Trecate, G.F.; Karimi, H.; Fulwani, D.; Lu, X.; Guerrero, J.M. Review on Control of DC Microgrids and Multiple Microgrid Clusters. *IEEE J. Emerg. Sel. Top. Power Electron.* **2017**, *5*, 928–948.
7. Jin, Z.; Sulligoi, G.; Cuzner, R.; Meng, L.; Vasquez, J.C.; Guerrero, J.M. Next-Generation Shipboard DC Power System: Introduction Smart Grid and dc Microgrid Technologies into Maritime Electrical Netowrks. *IEEE Electrif. Mag.* **2016**, *4*, 45–57. [CrossRef]
8. Emadi, A.; Williamson, S.S.; Khaligh, A. Power Electronics Intensive Solutions for Advanced Electric, Hybrid Electric, and Fuel Cell Vehicular Power Systems. *IEEE Trans. Power Electron.* **2006**, *21*, 567–577. [CrossRef]
9. Dragičević, T.; Lu, X.; Vasquez, J.C.; Guerrero, J.M. DC Microgrids—Part I: A Review of Control Strategies and Stabilization Techniques. *IEEE Trans. Power Electron.* **2016**, *31*, 4876–4891.
10. Dragičević, T.; Lu, X.; Vasquez, J.C.; Guerrero, J.M. DC Microgrids—Part II: A Review of Power Architectures, Applications, and Standardization Issues. *IEEE Trans. Power Electron.* **2016**, *31*, 3528–3549. [CrossRef]
11. Kwasinski, A.; Onwuchekwa, C.N. Dynamic behavior and stabilization of DC microgrids with instantaneous constant-power loads. *IEEE Trans. Power Electron.* **2011**, *26*, 822–834. [CrossRef]
12. Emadi, A.; Khaligh, A.; Rivetta, C.H.; Williamson, G.A. Constant power loads and negative impedance instability in automotive systems: Definition, modeling, stability and control of power electronic converters and motor drives. *IEEE Trans. Veh. Technol.* **2006**, *55*, 1112–1125. [CrossRef]
13. Rahimi, A.M.; Emadi, A. Active Damping in DC/DC Power Electronic Converters: A Novel Method to Overcome the Problems of Constant Power Loads. *IEEE Trans. Ind. Electron.* **2009**, *56*, 1428–1439. [CrossRef]
14. Rahimi, A.M.; Williamson, G.A.; Emadi, A. Loop-Cancellation Technique: A Novel Nonlinear Feedback to Overcome the Destabilizing Effect of Constant-Power Loads. *IEEE Trans. Veh. Technol.* **2010**, *59*, 650–661. [CrossRef]
15. Bosich, D.; Giadrossi, G.; Sulligoi, G. Voltage control solutions to face the CPL instability in MVDC shipboard power systems. In Proceedings of the AEIT Annual Conference 2014, Trieste, Italy, 18–19 September 2014.
16. Sulligoi, G.; Bosich, D.; Giadrossi, G.; Zhu, L.; Cupelli, M.; Monti, A. Multiconverter Medium Voltage DC Power Systems on Ships: Constant-Power Loads Instability Solution using Linearization via State Feedback Control. *IEEE Trans. Smart Grid* **2014**, *5*, 2543–2552. [CrossRef]
17. Cupelli, M.; Ponci, F.; Sulligoi, G.; Vicenzutti, A.; Edrington, C.S.; El-Mezyani, T.; Monti, A. Power Flow Control and Network Stability in an All-Electric Ship. *Proc. IEEE* **2015**, *103*, 2355–2380. [CrossRef]
18. Bosich, D.; Sulligoi, G.; Mocanu, E.; Gibescu, M. Medium Voltage DC Power Systems on Ships: An Offline Parameter Estimation for Tuning the Controllers' Linearizing Function. *IEEE Trans. Energy Convers.* **2017**, *32*, 748–758. [CrossRef]
19. Hossain, E.; Perez, R.; Nasiri, A.; Padmanaban, S. A Comprehensive Review on Constant Power Loads Compensation Techniques. *IEEE Access* **2018**, *6*, 33285–33305. [CrossRef]
20. Su, M.; Liu, Z.; Sun, Y.; Han, H.; Hou, X. Stability Analysis and Stabilization Methods of DC Microgrid with Multiple Parallel-Connected DC–DC Converters Loaded by CPLs. *IEEE Trans. Smart Grid* **2018**, *9*, 132–142. [CrossRef]
21. Riccobono, A.; Santi, E. Comprehensive Review of Stability Criteria for DC Power Distribution Systems. *IEEE Trans. Ind. Appl.* **2014**, *50*, 3525–3535. [CrossRef]
22. Riccobono, A.; Cupelli, M.; Monti, A.; Santi, E.; Roinila, T.; Abdollahi, H.; Arrua, S.; Dougal, R.A. Stability of Shipboard DC Power Distribution: Online Impedance-Based Systems Methods. *IEEE Electrif. Mag.* **2017**, *5*, 55–67. [CrossRef]
23. Javaid, U.; Freijedo, F.D.; Dujic, D.; van der Merwe, W. Dynamic Assessment of Source–Load Interactions in Marine MVDC Distribution. *IEEE Trans. Ind. Electron.* **2017**, *64*, 4372–4381. [CrossRef]
24. Belkhayat, M.; Cooley, R.; Witulski, A. Large Signal Stability Criteria for Distributed Systems with Constant Power Loads. In Proceedings of the 26th IEEE Annual Power Electronics Specialists Conference, Atlanta, GA, USA, 18–22 June 1995; pp. 1333–1338.
25. Griffo, A.; Wang, J.; Howe, D. Large Signal Stability Analysis of DC Power Systems with Constant Power Loads. In Proceedings of the IEEE Vehicle Power and Propulsion Conference (VPPC), Harbin, China, 3–5 September 2008; pp. 1–6.
26. Herrera, L.; Zhang, W.; Wang, J. Stability Analysis and Controller Design of DC Microgrids with Constant Power Loads. *IEEE Trans. Smart Grid* **2017**, *8*, 881–888.

27. Aboushady, A.A.; Ahmed, K.H.; Finney, S.J.; Williams, B.W. Lyapunov-based high-performance controller for modular resonant DC/DC converters for medium-voltage DC grids. *IET Power Electron.* **2017**, *10*, 2055–2064. [CrossRef]
28. Mukherjee, N.; Strickland, D. Control of Cascaded DC–DC Converter-Based Hybrid Battery Energy Storage Systems—Part II: Lyapunov Approach. *IEEE Trans. Ind. Electron.* **2016**, *63*, 3050–3059. [CrossRef]
29. Kabalan, M.; Singh, P.; Niebur, D. Large Signal Lyapunov-Based Stability Studies in Microgrids: A Review. *IEEE Trans. Smart Grid* **2017**, *8*, 2287–2295. [CrossRef]
30. Bosich, D.; Gibescu, M.; Sulligoi, G. Large-signal stability analysis of two power converters solutions for DC shipboard microgrid. In Proceedings of the 2017 IEEE Second International Conference on DC Microgrids (ICDCM), Nuremburg, Germany, 27–29 June 2017; pp. 125–132.
31. Sulligoi, G.; Bosich, D.; Giadrossi, G. Linearizing voltage control of MVDC power systems feeding constant power loads: Stability analysis under saturation. In Proceedings of the 2013 IEEE Power & Energy Society General Meeting, Vancouver, BC, Canada, 21–25 July 2013.
32. Grillo, S.; Musolino, V.; Sulligoi, G.; Tironi, E. Stability enhancement in DC distribution systems with constant power controlled converters. In Proceedings of the IEEE 15th International Conference on Harmonics and Quality of Power (ICHQP), Hong Kong, China, 17–20 June 2012; pp. 848–854.
33. Bosich, D.; Giadrossi, G.; Sulligoi, G.; Grillo, S.; Tironi, E. More Electric Vehicles DC Power Systems: A Large Signal Stability Analysis in presence of CPLs fed by Floating Supply Voltage. In Proceedings of the IEEE International Electric Vehicle Conference (IEVC), Florence, Italy, 17–19 December 2014; pp. 1–6.
34. Chen, M.; Rincón-Mora, G.A. Accurate Electrical Battery model capable of predicting runtime and I-V performance. *IEEE Trans. Energy Convers.* **2006**, *21*, 504–511. [CrossRef]
35. Rahimi, A.M.; Emadi, A. An Analytical Investigation of DC/DC Power Electronic Converters with Constant Power Loads in Vehicular Power Systems. *IEEE Trans. Veh. Technol.* **2009**, *58*, 2689–2702. [CrossRef]
36. Kuznestov, Y.A. *Elements of Applied Bifurcation Theory*, 3rd ed.; Springer: New York, NY, USA, 2004.
37. Arcidiacono, V.; Monti, A.; Sulligoi, G. Generation control system for improving design and stability of medium-voltage DC power systems on ships. *IET Electr. Syst. Transp.* **2012**, *2*, 158–167. [CrossRef]
38. Khalil, H.K. *Nonlinear Systems*; Prentice Hall: Upper Saddle River, NJ, USA, 1992.

applied sciences

MDPI

Article

Multidimensional Optimal Droop Control for DC Microgrids in Military Applications

Kaitlyn J. Bunker [1,2,*], Michael D. Cook [3], Wayne W. Weaver [1] and Gordon G. Parker [3]

1 Department of Electrical & Computer Engineering, Michigan Technological University, Houghton, MI 49931, USA; wwweaver@mtu.edu
2 Rocky Mountain Institute, Boulder, CO 80302, USA
3 Depatment of Mechanical Engineering & Engineering Mechanics, Michigan Technological University, Houghton, MI 49931, USA; mdcook@mtu.edu (M.D.C.); ggparker@mtu.edu (G.G.P.)
* Correspondence: kbunker@rmi.org; Tel.: +1-734-787-2068

check for updates

Received: 30 September 2018; Accepted: 16 October 2018; Published: 18 October 2018

Abstract: Reliability is a key consideration when microgrid technology is implemented in military applications. Droop control provides a simple option without requiring communication between microgrid components, increasing the control system reliability. However, traditional droop control does not allow the microgrid to utilize much of the power available from a solar resource. This paper applies an optimal multidimensional droop control strategy for a solar resource connected in a microgrid at a military patrol base. Simulation and hardware-in-the-loop experiments of a sample microgrid show that much more power from the solar resource can be utilized, while maintaining the system's bus voltage around a nominal value, and still avoiding the need for communication between the various components.

Keywords: solar; microgrid; droop control; patrol base

1. Introduction

The microgrid concept is increasingly being investigated and implemented in military contexts. Research has been conducted on the topic of implementing microgrids at fixed military installations within the United States [1,2]. This paper will focus on the application of microgrid technologies in support of overseas military expeditionary or contingency operations. Specific examples include intermediate staging bases (ISB), forward operating bases (FOB), combat outposts (COP), observation posts (OP), and patrol bases (PB) [3–5]. These may be permanent or temporary bases used by the military while on assignment in foreign countries. This paper will include a demonstration of a COP with a size of 20–40 people, so this type of microgrid will be considered here.

There are several challenges when considering microgrids for military applications. First, the COPs may be somewhat temporary, so equipment that can be easily transported to the location, then moved to future locations is important to consider for equipment selection, but the energy cost of transportation is outside the scope of this paper. Reliability is especially important in military microgrids, since many lives depend on the consistent operation of the electrical equipment at the COP [6]. Minimizing the use of fuel is also critical, as transporting fuel to the COP can cost $400 per gallon [7], and can be dangerous, resulting in the loss of one life for every 24 fuel convoys in Iraq and Afghanistan in 2007 [8].

Including renewable resources can help reduce the amount of fuel needed, and energy storage can be included to increase the reliability of the system. Assessing the amount of storage needed for energy surety has been previously studied [9,10]. Using the improved droop control methods presented in this paper can allow a military microgrid to utilize as much renewable resources as possible, without relying on a communication link between the microgrid components—this eliminates the electrical system communications as a single source of failure that would be especially vulnerable in a military setting.

Previous work has demonstrated the use of multidimensional [11] optimal [12] droop control with wind resources, in a residential microgrid setting. This paper expands upon previous work, applying multidimensional optimal droop control with solar resources. A load model for a typical patrol base is presented, and used in simulation of a military microgrid example to demonstrate the proposed controller.

2. Load Modeling

In a military setting, the load on the microgrid is different than in a residential or commercial setting. A load model for a patrol base (PB) was developed with the help of an experienced member of the military [13]. First, three load tiers were identified to distinguish between loads that may be connected at a PB. Tier 1 includes loads that are critical to the mission, while Tier 2 includes loads that are important for extended operations. Tier 3 includes loads for improving quality of life and boosting morale. The loads in Tier 3 are important for maintaining the living conditions and overall attitudes for the PB members, however they could be disconnected during some time periods in order to ensure that Tier 1 and 2 loads are met. Table 1 lists some examples of equipment that might be included in each load tier.

The amount of load needed in each tier varies based on time (day vs. night) as well as on activity (patrol vs. non-patrol). Table 2 shows the estimated load for a 20–40 person PB in each of these scenarios.

Table 1. Equipment included in each load tier at a Patrol Base.

Load Tier	Equipment
Tier 1 (Mission Critical)	One to ten 80 W Panasonic Toughbooks Charging station for radios capable of recharging (30) 15 W radios One or two AC units One–Two TVs/projectors used as large monitors for the Blue Force Tracker Three to five 100 W bulbs (26 W CFL)
Tier 2 (Extended Operations)	Power tools and charging station if the platoon is equipped with vehicles ISR video feeds Portable lights powered via rechargeable battery pack One to three hotplates/coffee pots
Tier 3 (Morale Boosting)	One to three personal electronic devices (iPod, laptop, camera, etc.) 30–80 gallon hot water heater Three to five additional AC/heater units for berthing One to three flat panel TVs or projectors

Table 2. Amount of load in each tier based on time and activity.

Load Tier	Activity	Day	Night
Tier 1	Patrol	2–3 kW	2–3 KW
	Non-Patrol	1–2.5 kW	1–2.5 kW
Tier 2	Patrol	1–2 kW	1–2 KW
	Non-Patrol	2–3 kW	1–2 kW
Tier 3	Patrol	2–3 kW	2–3 KW
	Non-Patrol	0.5–1.5 kW	0.5–1 kW

The estimates shown in Table 2 are for the total load used in each different type of scenario. The actual load is stochastic and will vary, so it was modeled using a compound Poisson process. A Poisson process is defined as

$$P[N(t) = k] = e^{-\lambda t}\frac{(\lambda t)^k}{k!} \tag{1}$$

where λ is the rate parameter. The Poisson process $N(t)$ is then implemented in a compound Poisson process as

$$Y(t) = \sum_{i=1}^{N}(t)D_i \tag{2}$$

where D_i are independent and identically distributed (iid) random variables [11]. A compound Poisson process is an appropriate method for approximating the distribution of electrical loads, since they are discrete, independent random variables [15].

Along with the load changing stochastically within each time and activity, the operation of the PB will also vary. In any combat environment, varying patrol patterns (route, length of patrol, time of departure, etc.) are used in order to limit the enemy's ability to execute a coordinated attack. Therefore, this load model also randomizes the number of patrols per day, their length, and the length of rest during both day and night. Some days may then include no patrols, to allow for debriefing or rest. Other days may include a high number of patrols, designed to limit the enemy's ability to operate. An example patrol schedule for two squads during one day and one night is shown in Table 3; the patrol schedule is randomized for other days and nights.

Table 3. Example patrol schedule.

Day 1 Patrol Schedule	
0725–0800	Squad 1
1110–1325	Squad 2
1610–1800	Squad 1
Night 1 Patrol Schedule	
1910–1955	Squad 1
2140–2225	Squad 2
2340–0050	Squad 1
0300–0400	Squad 2

A load model incorporating all of these aspects was created using Matlab. Figure 1 shows the load at a PB over one day, based on a changing patrol schedule and using a compound Poisson process to randomize the load. Each load tier is displayed in the figure. This load model is implemented in simulation and HIL in the following sections.

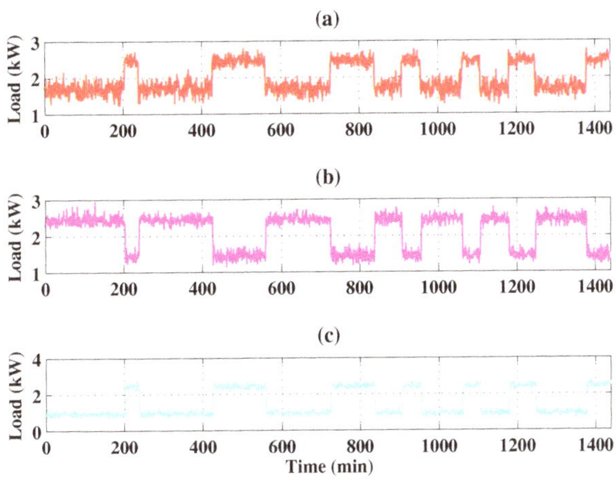

Figure 1. Stochastic load at a Patrol Base microgrid for (**a**) Tier 1; (**b**) Tier 2; and (**c**) Tier 3 loads.

3. Source Modeling

For a military application, it is likely that solar would be included as a renewable resource, instead of wind. Like wind resources, solar resources are intermittent and depend on weather conditions. The accurate prediction of solar irradiance is currently an important research area [16].

For military applications, portability is a key consideration [17], and the use of solar resources fits this requirement. Previous research has been completed on the inclusion of solar resources in military forward operating bases [18]. In this section, the proposed optimal high dimension droop control will be applied with a solar resource; only applications with wind resources have been demonstrated in previous chapters.

Solar irradiance measurements are taken each minute by the National Renewable Energy Laboratory (NREL). These are recorded at their Golden, CO location and stored as part of the Baseline Measurement System (BMS). The data for 1 June 2012 is used in simulation and HIL in the following sections, and is shown in Figure 2 [19].

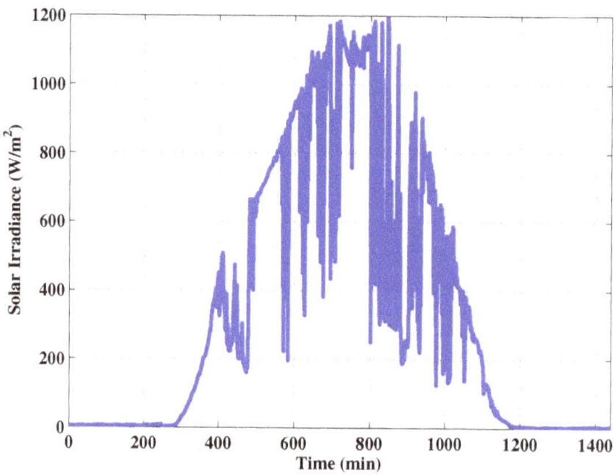

Figure 2. Solar irradiance measured each minute on 1 June 2012.

4. Simulation Results

The sample microgrid used in simulation for a PB is shown in Figure 3. Three sources are included; one is a solar resource, one is a conventional resource such as a diesel generator, and one is an energy storage source such as a battery. In the simulation, these sources are modeled as variable voltage sources, and droop control is implemented for each. The numeric values used for this microgrid model are shown in Table 4.

Table 4. Patrol base microgrid parameter values.

Component	Value	Unit
R_{1B}	0.1	Ω
L_{1B}	1	mH
R_{2B}	0.2	Ω
L_{2B}	2	mH
R_{3B}	0.12	Ω
L_{3B}	1	mH
C_{load}	1	mF
k_p	1	
k_i	10	

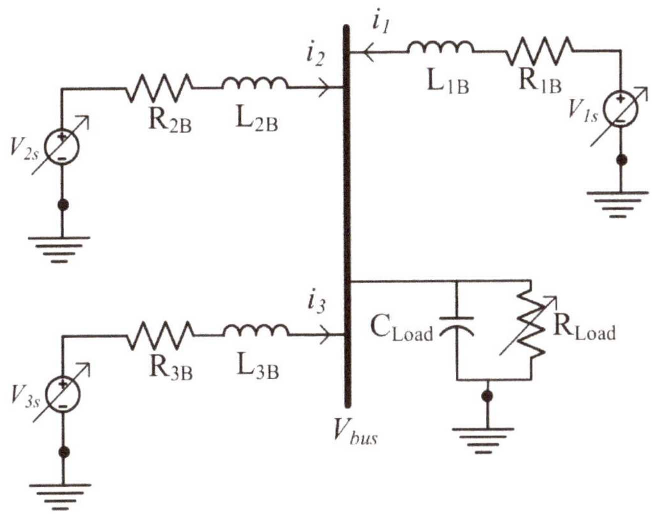

Figure 3. Example microgrid used in simulation for patrol base (PB) demonstration of droop control methods.

A simulation was completed to represent one full day at the PB. The power available from solar energy at each minute of the day is shown in Figure 4a. The data described in the previous section is used, assuming that there are 4 square meters of solar panels available. The required load at each minute is shown in Figure 4b, with the three load tiers described above added together to form the total required load.

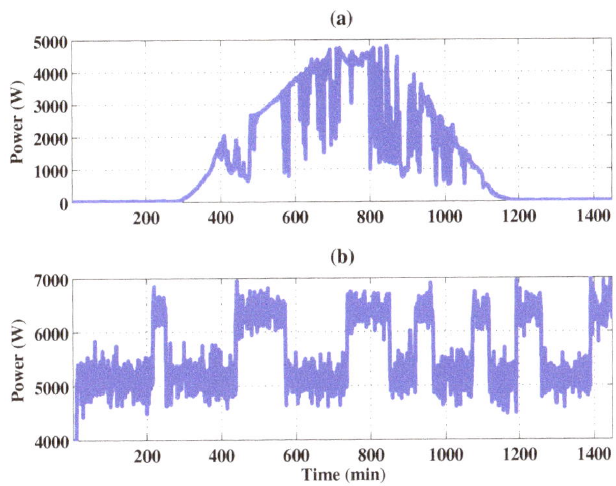

Figure 4. (**a**) Solar power available and (**b**) load profiles for simulated 1440 min period.

The system was first simulated using traditional, linear droop control for each of the three sources, where the reference current is defined as

$$i_{ref} = \frac{V_{ref} - V_{bus}}{R_d} \tag{3}$$

The droop control settings for each of the three sources are shown in Table 5.

Table 5. Traditional droop control settings for patrol base (PB) simulation.

Source	V_{ref}	R_d
Solar	300	2.5
Conventional	300	0.4
Storage	295	1.0

The power supplied by each of the three sources during the simulated one day period is shown in Figure 5. When it is sunny during the day, the solar source operates with traditional droop control, and does not use all of the available power. The conventional source is required to meet most of the load demand. The storage source supplies and absorbs power throughout the day as the load changes.

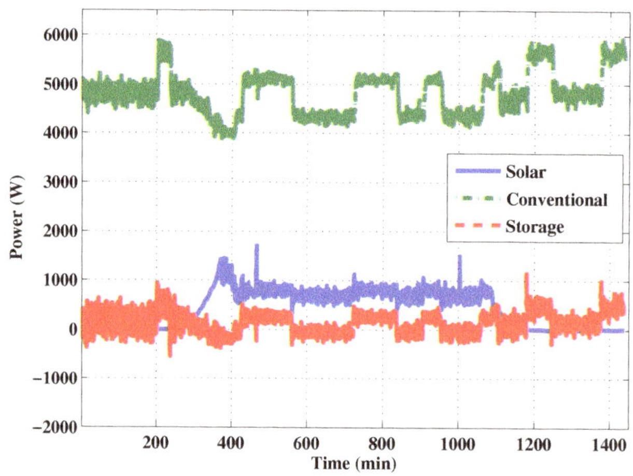

Figure 5. Power supplied by solar, conventional, and storage sources when traditional droop control is used.

The simulation was then repeated using optimal high dimension droop control for the solar resource. Its reference current is defined as [12]

$$i^*_{ref1} = \frac{-V_{bus} + \sqrt{4\hat{P}(s)R_{1B} + V_{bus}^2}}{2R_{1B}} \tag{4}$$

where s is the solar irradiance in W/m^2 multiplied by the 4 m^2 of panels. The conventional and storage sources kept the same droop settings as in Table 5 for the second simulation.

The power supplied by each of the three sources during the simulated one day period is shown in Figure 6. With optimal high dimension droop control, all of the available power from the solar resource is utilized. This means that the conventional source is required to provide less power to the system. The storage source changes during the sunny part of the day, and supplies power as needed during the night.

A comparison of Figures 5 and 6 shows that the use of traditional linear droop control limits the amount of power that can be utilized from the solar resource, while the use of optimal high dimension droop control allows all of the available power from that resource to be utilized. This also means that the conventional source is needed less, and the storage source is able to charge during times of high irradiance, and use that stored energy during cloudy periods or at night.

The bus voltage during each of the two simulations is shown in Figure 7. While the bus voltage varies more when optimal high dimension droop control is used, it stays well within a bound of 5% around the reference value of 300 V.

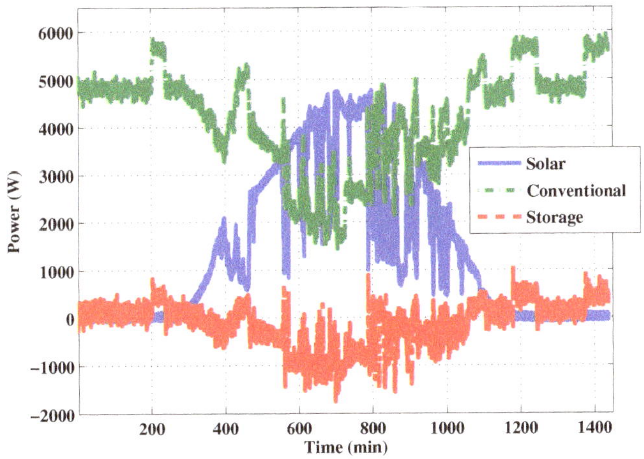

Figure 6. Power supplied by solar, conventional, and storage sources when optimal high dimension droop control is used.

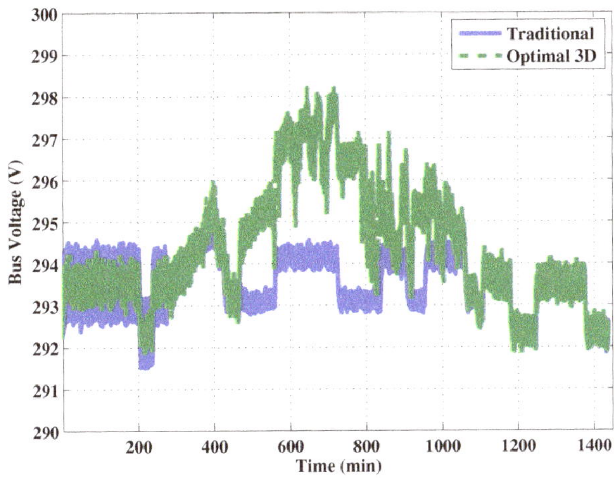

Figure 7. Bus voltage using traditional vs. optimal high dimension droop control.

5. Hardware-In-The-Loop Results

In order to validate the simulation results presented in the previous section, the example microgrid shown in Figure 8 was implemented using a hardware-in-the-loop (HIL) system, allowing the system to be emulated in software, and connected with real world hardware in real time [20]. Using an HIL approach allows the system to be built and tested virtually with new proposed control methods, while interfacing with real hardware [21].

The same component values were implemented in the simulation, and in a Typhoon HIL600 unit. The HIL experimental apparatus is shown in Figure 9. The Typhoon HIL600 has 32 channels of ±5 V analog output (AO) that can be mapped to data points in the HIL circuit. The microgrid control board then offsets and scales these signals to 0 V to +3.3 V that are read by the 12-bit analog to digital converters on all three DSPs. For this example, only 5 channels of analog signals are needed to implement the proposed control. The analog signals and the scaling from the HIL are shown in Table 6.

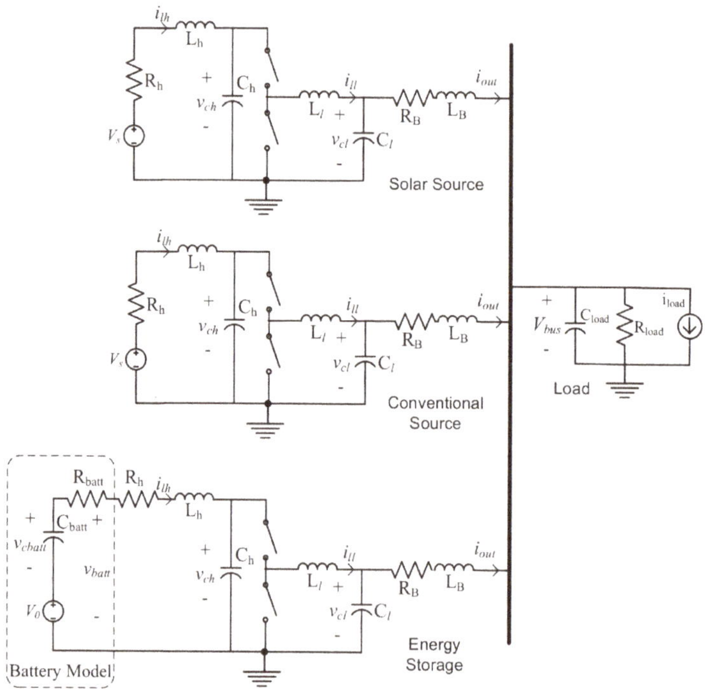

Figure 8. Example microgrid used in hardware-in-the-loop (HIL) for PB demonstration of droop control methods.

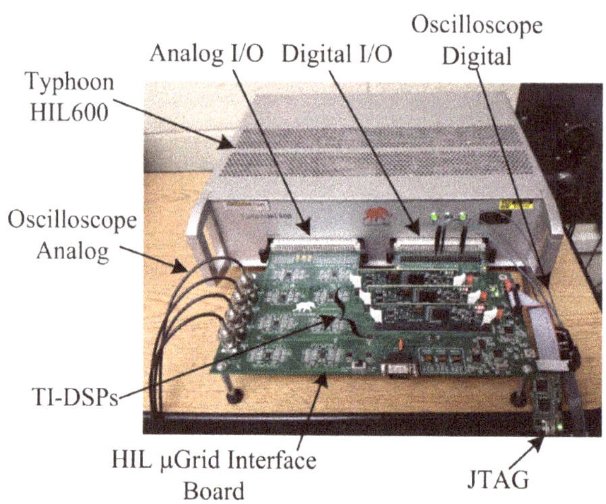

Figure 9. Typhoon HIL600 with microgrid control board and three TI-F28335 DSP ControlCards (two cards were used for this paper).

Table 6. Hardware-in-the-loop (HIL) Analog Output Configuration.

AO Channel	Model Parameter	Scaling
1	V_{bus}	100 V *per* 1 V_{dc}
2	I_1	10 A *per* 1 V_{dc}
3	I_2	10 A *per* 1 V_{dc}
4	I_3	10 A *per* 1 V_{dc}
5	$V_{1,ref}$	100 V *per* 1 V_{dc}
6	$V_{2,ref}$	100 V *per* 1 V_{dc}
7	$V_{3,ref}$	100 V *per* 1 V_{dc}

The microgrid control board has BNC connectors linked with the first four analog output channels that are convenient to connect to an oscilloscope. For this work, the first three channels were used to view the first three signals in Table 6, and capture the results in oscilloscope traces.

The controller for each source was implemented on a separate Texas Instruments F28335 DSP ContolCARD [22] programmed through the Embedded Coder toolbox in MATLAB/Simulink. The control card for Source 1 implements (4), while the card for Source 2 implements (3), each using a PID control loop. By using separate control cards for each of the two sources, decentralized control is ensured; each source uses only local information, and the proposed control method does not require a communication link between the sources and/or the other components in the microgrid.

This microgrid has the same topology as the one implemented in the simulation. However, for the HIL experiment, power electronics were included, along with parasitic inductances and capacitances. To represent the variable load, a controllable current source was included. This type of load is comparable to an inductive or motor load in an actual system. The numeric values used for the circuit parameters are shown in Tables 7 and 8.

Table 7. Source 1 and Source 2 parameter values.

Source 1			Source 2		
Component	Value	Unit	Component	Value	Unit
V_h	400	V	V_h	380	V
R_h	0.2	Ω	R_h	0.2	Ω
C_h	0.2	mF	C_h	0.2	mF
L_h	1.5	mH	L_h	1.5	mH
C_l	0.08	F	C_l	0.08	F
L_l	5	mH	L_l	5	mH
R_B	0.1	Ω	R_B	0.23	Ω
L_B	1.2	mH	L_B	0.8	mH
V_{ref}	300	V	V_{ref}	300	V
R_d	0.9		R_d	0.8	

Table 8. Energy storage parameter values.

Component	Value	Unit
C_h	1	mF
L_h	0.2	mH
C_l	0.08	F
L_l	1	mH
R_B	1	Ω
L_B	1.2	mH
V_{ref}	293	V
R_d	1.5	

The system was implemented on a Typhoon HIL600 unit using distributed control through Texas Instruments F28335 DSP ContolCARDs. Each of the three sources is controlled using a separate TI card, using only local sensors. The circuit schematic used in the Typhoon software is shown in Figure 10.

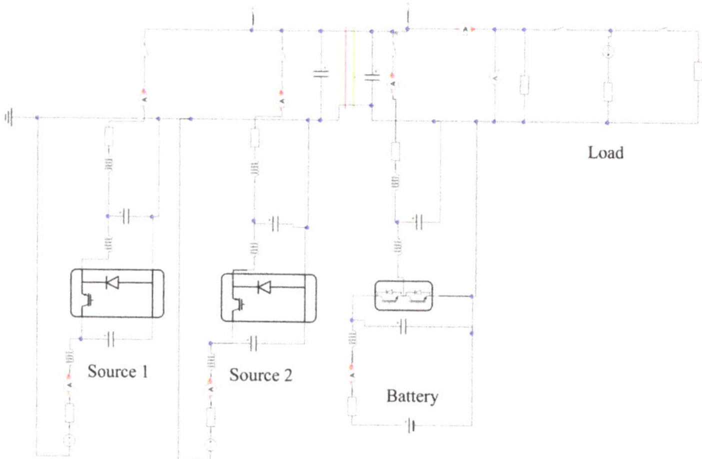

Figure 10. Hardware-in-the-loop schematic for PB microgrid example.

The controllers implemented for each of the three sources are shown in Figures 11–13. Sources 2 and 3, representing a conventional and energy storage source, respectively, are controlled using traditional linear droop control. Source 1, representing a solar source, is controlled using optimal high dimension droop control as in (4).

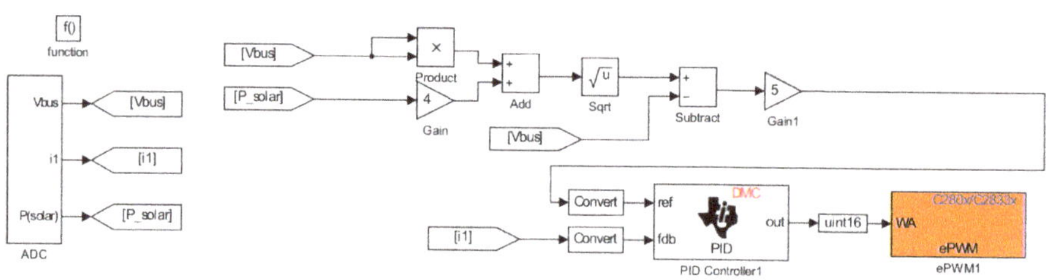

Figure 11. HIL controller for Source 1 (solar)–optimal high dimension droop.

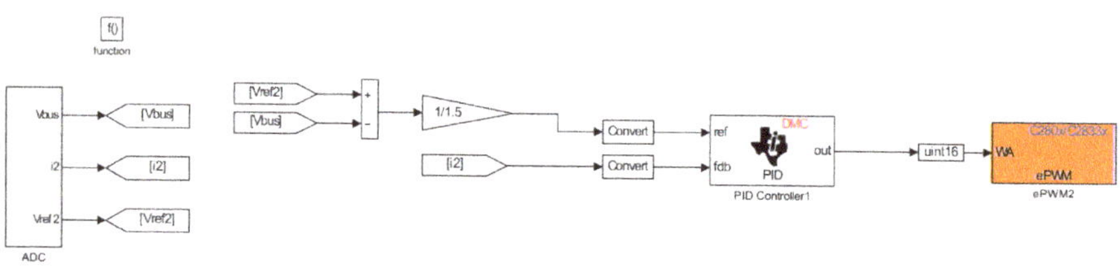

Figure 12. HIL controller for Source 2 (conventional)–traditional droop.

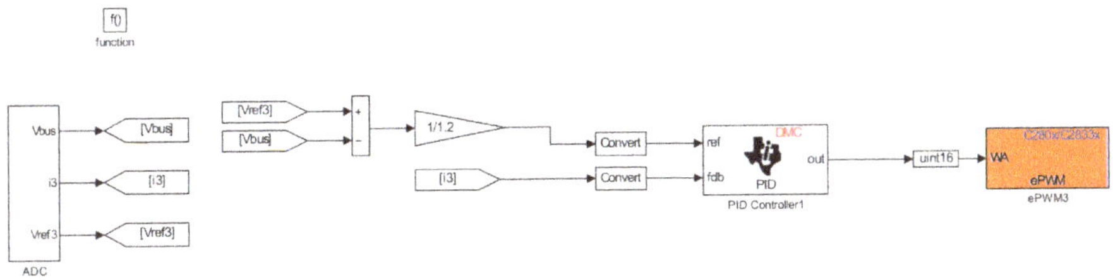

Figure 13. HIL controller for Source 3 (storage)–traditional droop.

Due to the limitations of the HIL system, a short overall time period was used, and the per-minute solar and load information was scaled. The profiles for power available from the solar resource and required load current are shown in Figure 14. The solar irradiance data is taken from the NREL BMS measurements from 1 June 2012, starting at 11:00 a.m.

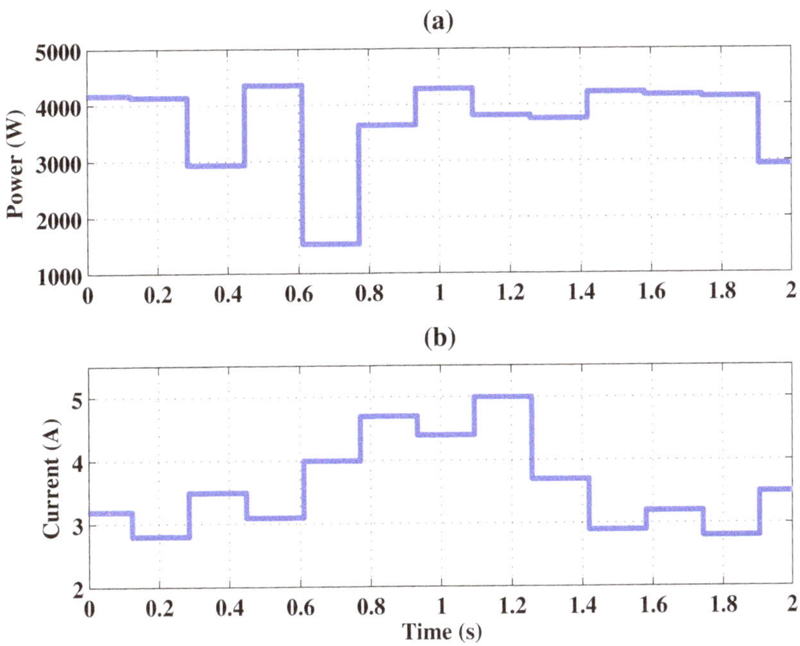

Figure 14. (**a**) Solar power available and (**b**) load profiles for HIL implementation.

The results of implementing these profiles in the HIL system are shown in Figure 15. The bus voltage is shown, along with the current supplied from each of the three sources—all three use the same zero point on the oscilloscope trace. As the power available from the sun and the required load change, the conventional and storage sources change their output based on traditional linear droop control. Source 1 changes its output based on optimal high dimension droop control, where the reference current is determined by (4).

In order to verify that the proposed controller is operating as desired, data from the oscilloscope trace in Figure 15 was imported and plotted using MATLAB. The bus voltage and solar power available were used in (4) to calculate the reference current that Source 1 should be supplying. This reference current is plotted with the actual Source 1 current in Figure 16. The source output does match the desired reference value calculated using optimal high dimension droop control.

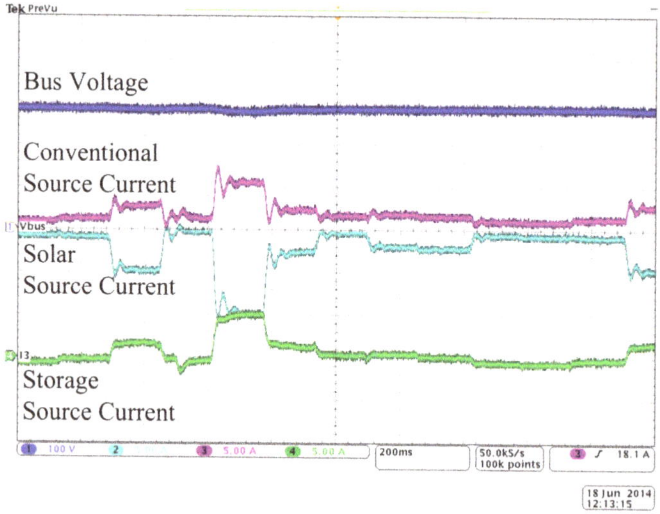

Figure 15. Hardware-in-the-loop results with optimal high dimension droop control: Ch1 Bus Voltage; Ch2 Source 1 Current; Ch3 Source 2 Current; Ch4 Source 3 Current.

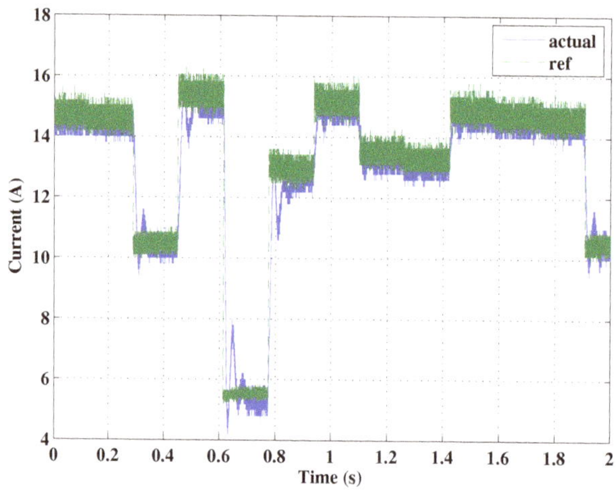

Figure 16. Reference and actual current supplied by solar resource using optimal high dimension droop control.

6. Conclusions

The results presented in this paper show that the droop control relationship for a source in a DC microgrid can be optimized to meet a given objective. An example microgrid was simulated and demonstrated through HIL, and the results show that a droop control relationship can be chosen to allow the power supplied by a source to match the power available from the solar PV resource.

This control method retains the advantages of traditional droop control, and does not require a communication link between the system components. It also allows all of the power available from solar PV resource to be utilized, which is an improvement over traditional droop control, and over high dimension droop control using a plane. The droop control surface is optimized for general operation of the system—it is not necessary to know the expected solar irradiance and load profiles in order to determine the surface shape.

Author Contributions: Conceptualization, K.J.B., M.D.C., W.W.W. and G.G.P.; Formal analysis, K.J.B.; Investigation, K.J.B. and M.D.C.; Methodology, K.J.B., M.D.C., W.W.W. and G.G.P.; Supervision, W.W.W.; Writing—original draft, K.J.B.; Writing—review & editing, M.D.C., W.W.W. and G.G.P.

Funding: This work was supported by a National Science Foundation Graduate Research Fellowship (DGE-1051031), and the Army Research Laboratory (W911NF-13-2-0024).

Conflicts of Interest: The authors declare no conflict of interest.

References

1. Johnson, M.D.; Ducey, R.A. Overview of U.S. Army microgrid efforts at fixed installations. In Proceedings of the 2011 IEEE Power and Energy Society General Meeting, Detroit, MI, USA, 24–29 July 2011.
2. Skowronska-Kurec, A.G.; Eick, S.T.; Kallio, E.T. Demonstration of microgrid technology at a military installation. In Proceedings of the 2012 IEEE Power and Energy Society General Meeting, San Diego, CA, USA, 22–26 July 2012.
3. United States Army Corps of Engineers. Base Camp Development in the Theater of Operations. EP 1105-3-1. 2009. Available online: https://www.publications.usace.army.mil/Portals/76/Publications/EngineerPamphlets/EP_1105-3-1.pdf (accessed on 16 October 2018).
4. United States Army; United States Marine Corps. *Base Camps*; ATP 3-37.10/MCRP 3-17.7N; CreateSpace Independent Publishing Platform: Scotts Valley, CA, USA, 2013.
5. United States Army. Infantry Small-Unit Mountain Operations. ATTP 3-21.50. 2011. Available online: https://www.adlibris.com/fi/kirja/infantry-small-unit-mountain-operations-attp-3-2150-9781480119932 (accessed on 16 October 2018).
6. Hartranft, T.J.; Yeboah, F.; Grady, D.; Ducey, R. Energy Security and Independence for Military Installations: Candidate Mission-Focused Vision and Policy Measures. In Proceedings of the 1st Army Installation Energy Security and Independence Conference, Long Beach, CA, USA, 27–30 June 2007; pp. 1–28.
7. Rosenthal, U.S. Military Orders Less Dependence on Fossil Fuels. 2010. Available online: http://www.nytimes.com/2010/10/05/science/earth/05fossil.html (accessed on 16 October 2018).
8. Eady, D.S.; Siegel, S.B.; Bell, R.S.; Dicke, S.H. *Sustain the Mission Project: Casualty Factors for Fuel and Water Resupply Convoys*; Army Environmental Policy Institute: Herndon, VI, USA, 2007; pp. 1–28.
9. Vallem, M.R.; Jensen, D.; Mitra, J. Reliability evaluation and need based storage assessment for surety microgrids. In Proceedings of the 2006 38th North American Power Symposium, Carbondale, IL, USA, 17–19 September 2006; pp. 29–33.
10. Mitra, J.; Vallem, M.R. Determination of storage required to meet reliability guarantees on island-capable microgrids with intermittent sources. *IEEE Trans. Power Syst.* **2012**, *27*, 2360–2367. [CrossRef]
11. Bunker, K.J.; Weaver, W.W. Multidimensional droop control for wind resources in dc microgrids. *IET Gener. Trans. Distrib.* **2017**, *11*, 657–664. [CrossRef]
12. Bunker, K.J.; Weaver, W.W. Multidimensional droop control for wind resources in dc microgrids. *Energies* **2018**, *11*, 1–20. [CrossRef]
13. Cook, M. Preliminary report on ARL project. In Proceedings of the 6th Northumbria International Conference on Performance Measurement in Libraries and Information Services, Northumbria, UK, 22–25 July 2013; pp. 1–10.
14. Bowers, N.L.; Gerger, H.U.; Hickman, J.C.; Jones, D.A.; Nesbitt, C.J. *Actuarial Mathematics*; Society of Actuaries: Schaumburg, IL, USA, 2007.
15. Hu, Z.; Wang, X. A probabilistic load flow method considering branch outages. *IEEE Trans. Power Syst.* **2006**, *21*, 507–514. [CrossRef]
16. Salcedo-Sanz, S.; Casanova-Mateo, C.; Munoz-Mari, J.; Camps-Valls, G. Prediction of daily global solar irradiation using temporal gaussian processes. *IEEE Geosci. Remote Sens. Lett.* **2014**, *11*, 1936–1940. [CrossRef]
17. Trautz, K.M.; Jenkins, P.P.; Walters, R.J.; Scheiman, D.; Hoheisel, R.; Tatvarti, R.; Chan, R.; Miyamoto, H.; Adam, J.G.J.; Elarde, V.C.; et al. Mobile solar power. *IEEE J. Photovolt.* **2013**, *3*, 535–541. [CrossRef]

18. Tyner, J.; Coates, M.; Holloway, D.; Goldsmith, K.; Daniels, C.; Vranicar, T.; Roling, J.; Jensen, D.; Mundy, A.; Peterson, B. The design of a portable and deployable solar energy system for deployed military applications. In Proceedings of the Systems and Information Engineering Design Symposium (SIEDS), Charlottesville, VI, USA, 29 April 2011; pp. 50–53.
19. National Renewable Energy Laboratory. Solar Radiation Research Laboratory Historical Data. 2014. Available online: http://www.nrel.gov/midc/srrl_bms/ (accessed on 16 October 2018).
20. Ivanovic, Z.R.; Adzic, E.M.; Vekic, M.S.; Grabic, S.U.; Celanovic, N.L.; Katic, V.A. HIL evaluation of power flow control strategies for energy storage connected to smart grid under unbalanced conditions. *IEEE Trans. Power Electron.* **2012**, *27*, 4699–4710. [CrossRef]
21. Yousefpoor, N.; Azidehak, A.; Bhattacharya, S.; Parkhideh, B.; Celanovic, I.; Genic, A. Real-time hardware-in-the-loop simulation of convertible static transmission controller for transmission grid management. In Proceedings of the 2013 IEEE 14th Workshop on Control and Modeling for Power Electronics (COMPEL), Cantabria, Spain, 23–26 June 2013; pp. 1–8.
22. Texas Instruments. TMS320F28335 Controlcard. 2014. Available online: http://www.ti.com/tool/tmdscncd28335 (accessed on 16 October 2018).

applied sciences

MDPI

Article

Energy Management Strategy for Rural Communities' DC Micro Grid Power System Structure with Maximum Penetration of Renewable Energy Sources

Maheswaran Gunasekaran [1], Hidayathullah Mohamed Ismail [1], Bharatiraja Chokkalingam [1], Lucian Mihet-Popa [2,*] and Sanjeevikumar Padmanaban [3]

1 Department of Electrical and Electronics Engineering, SRM University, Chennai 603 203, India; maheshvaraneee@gmail.com (M.G.); hidayathullahm@gmail.com (H.M.I.); bharatiraja@gmail.com (B.C.)

2 Faculty of Engineering, Østfold University College, Kobberslagerstredet 5, 1671 Kråkeroy-Fredrikstad, Norway

3 Department of Energy Technology, Aalborg University, Esbjerg 6700, Denmark; san@et.aau.dk

* Correspondence: lucian.mihet@hiof.no; Tel.: +47-922-713-53

Received: 19 January 2018; Accepted: 12 March 2018; Published: 8 April 2018

check for updates

Abstract: The AC and DC power system structures need to be modernized to meet consumer demands. DC microgrids are suitably admired due to their high efficiency, consistency, reliability, and load sharing performance, when interconnected to DC renewable and storage sources. The main control objective for any DC microgrid is providing proper load–power balancing based on the Distributed Generator (DG) sources. Due to the intermittent nature of renewable energy sources, batteries play an important role in load–power balancing in a DC microgrid. The existing energy management strategy may be able to meet the load demand. However, that technique is not suitable forrural communities' power system structure. This research offers an energy management strategy (EMS) for a DC microgrid to supply power to rural communities with solar, wind, fuel cell, and batteries as input sources. The proposed EMS performs the load–power balancing between each source (renewable and storage) in a DC microgrid for dynamic load variation. Here, the EMS handles two battery sources (one is used to deliver power to the priority load, and the other is utilized in the common DC bus) to meet the required demand. The proposed EMS is capable of handling load–power balancing using renewable energy sources with less consumption of non- conventional energy sources (such as a diesel generator). The performance of the system is analyzed based on different operating conditions of the input sources. The MATLAB/Simulink simulation model for the proposed DC microgrid with their EMS control system is developed and investigated, and their results are tabulated under different input and load conditions. The proposed EMS is verified through a laboratory real-time DC microgrid experimental setup, and the results are discussed.

Keywords: energy management strategy (EMS); distributed generator (DG); state of charge (SoC); circuit breaker (CB); standard test condition (STC); photovoltaic (PV)

1. Introduction

Today, the world population is drastically increasing, hence the consumption of electrical energy is increasing accordingly. Since 75% of people are residents of the metropolitan areas and 25% of people live in remote villages, the majority of the power supplied to these areas is generated using fossil fuels rather than the renewable energy sources. Hence, the usage of fossil fuelswill result in 60% environmental depletion. Today, many countries have started to generate power through non-conventional energy sources compared to conventional energy sources. Considerable investments in distributed energy resources have taken place even during and after the financial crisis affecting the world economy from 2008 to 2012. As per an international energy agency report on world energy investment in 2017, China, the United States, Europe, India, Russia, and South Africa have invested a considerable amount to harvest renewable energies in comparison to other countries [1]. According to the international energy investment report [2], many countries have started investing much in solar and wind rather than other renewable energy sources.

The best solution to overcome environmental depletion is by harvesting a renewable energy–sourced microgrid. Most of the non-conventional energy sources are synchronized with the utility or load through power converter interfaces, which constitutes a microgrid. Microgrids are sub-categorized into three types: AC, DC, and hybrid [3]. AC microgrids have the benefit of using power from the utility.However, an AC microgrid needs a relatively complex controller for synchronizing the system and importing and exporting power while maintaining system stability. The DC microgrid has certain advantages with many features. On the generation side, some of the renewable energy sources, generating power in a DC microgrid include fuel cells and photovoltaic and storage systems, whereas on the load side, modern electronic loads like computers, telephone base stations, vehicle charging stations, and telecommunication applications require DC power. Both the AC and DC loads will be fulfilled by a hybrid microgrid. The main advantage of a microgrid is having a less transmission loss, electrification of remote villages, improving power system stability by supporting reactive power balance in the utility grid, and high reliability with energy storage devices to sustain power balance among the generation and demand. The total energy conversion loss of AC to DC to supply DC loads is approximately 10–25% [4,5]. To resolve this issue, the DC microgrid feeds the load directly, which will reduce the conversion loss [6,7]. For remote village communities, the DC microgrid is the most preferred power supplying utility because the main utility grid may not electrify these village communities because of transmission issues. Therefore, the residential, lighting and commercial loads will be met by the DC microgrid with the help of a battery [8–11]. The battery energy storage system will be important in the DC microgrid because the batteries are charged during non-peak load condition and discharged during peak load condition with some fluctuations in the power generated by the renewable energy sources [11–14]. There are several research areas on the subject of the microgrid, of which the demand side management (DSM) is the main focus area. It will adjust the demand with respect to the consumers' requirements with respect to time (kW/h). Therefore, the utility can reduce the peak load and reduce congestion in the system.

This system will decrease the difference between the peak load and total generation at peak time periods [15–17]. The transition from a minimum load to maximum load approachingthe compress load curve is a great advantage of demand-side management in microgrid applications. The energy storage has different ramp rates, which are used to reduce the deviation in the voltage and improve stability [15]. Biomass is one of the non-conventional energy sources that generate power in small capacity and that is also integrated into the DC microgrid to supply power to remote communities [16]. However, by integrating the various sources into the grid, protection has to be considered for the DC microgrid, which is discussed in [17,18]. The depth penetration of renewable energy sources into the microgrid and EMS when considering the intermittent of photovoltaic systems is proposed in [19,20]. The main control function of EMS in any DC microgrid structure is to handle the load–power balancing. It should be reliable and cost-effective based on the power obtained from the distributed generation [18–20]. A conventional EMS has some ability to meet the above-mentioned problems for the DC microgrid structure. A rural community microgrid design and a review of the various architectures in some countries are discussed in [21,22]. Power quality is one of the issues in hybrid and AC microgrids [21–26]. This research work will focus on the EMS for a DC microgrid to supply remote village communities. The proposed EMS overcomes the drawbacks of the conventional system by load–power balancing between each source (renewable and storage) in a DC microgrid for dynamic load variation and reduces the consumption of non-conventional energy sources (such as a diesel generator). The general concept of the microgrid, the load profile of the village communities, and the functional block diagram of the proposed DC microgrid are discussed in Section 2. In Section 3, the mathematical modeling of different renewable energy sources is discussed. In Section 4, the proposed EMS is discussed. The simulation studies and experimental results are discussed in Sections 5 and 6. Finally, the DC microgrid system is concluded in Section 7.

2. Power System Network

The power system network is comprised of generation, transmission, and distribution. In generation, all the power generation plants, like thermal, hydro, nuclear etc., are connected in parallel. The generated power is stepped up and transmitted to reduce transmission loss. Finally, in the distribution stage, the power is supplied to different types of consumers with respect to their requirements. Figure 1 shows the schematic structure of the power system network. There are two thermal units, two hydro, and two nuclear power generation units that are connected in parallel, which will be connected to the transmission network. In the second stage of the transmission network, the co-generation plants like sugar plants and cement plants are connected to the distribution bus, and finally, in the distribution network, the voltage is stepped down, and it will supply power to different consumers like industrial, commercial, and domestic consumers. The power fluctuation severely affects consumers in the distribution stage, which is limited to some extent through the use of the microgrid, which is supplied through renewable energy sources like wind, PV, fuel cells, and diesel power. The main advantage of the microgrid is that it is able to instantaneously meet any increase in load that cannot be reciprocated by thermal, hydro, and nuclear generating units. The liability of the system will improve, and the microgrid is a better solution for supplying power to a remote village community.

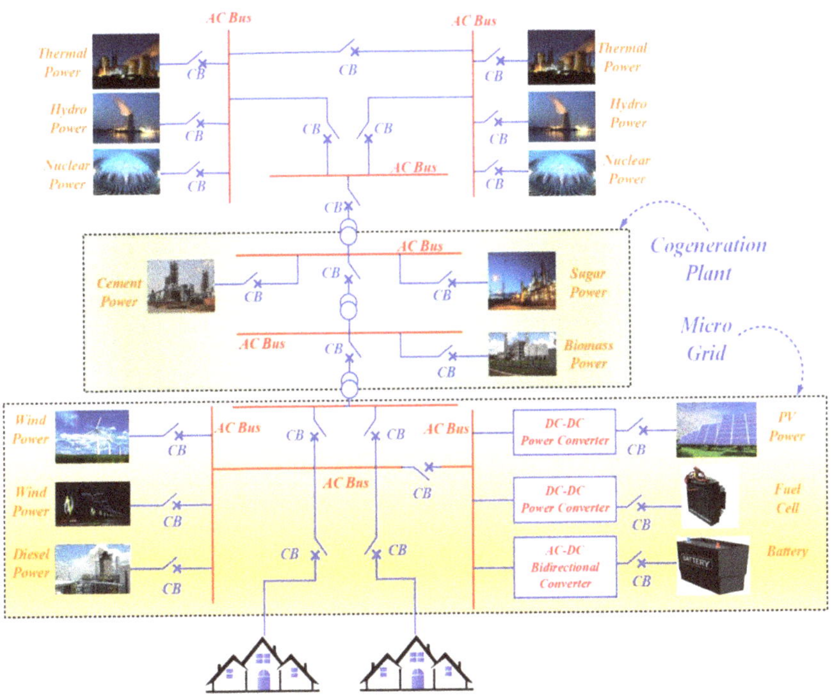

Figure 1. Power system network.

Proposed DC Microgrid Architecture for Rural Communities

The proposed DC microgrid architecture is for remote villages. In a village, the various loads such as commercial loads, agricultural vehicle loads, and priority load (street lighting) are considered. The power generation capacity of each sourceis 5 kW, as shown in Figure 2. The power generated in the fuel cell and PV is DC, but for wind it is AC, so the AC is converted into DC by means of a converter.

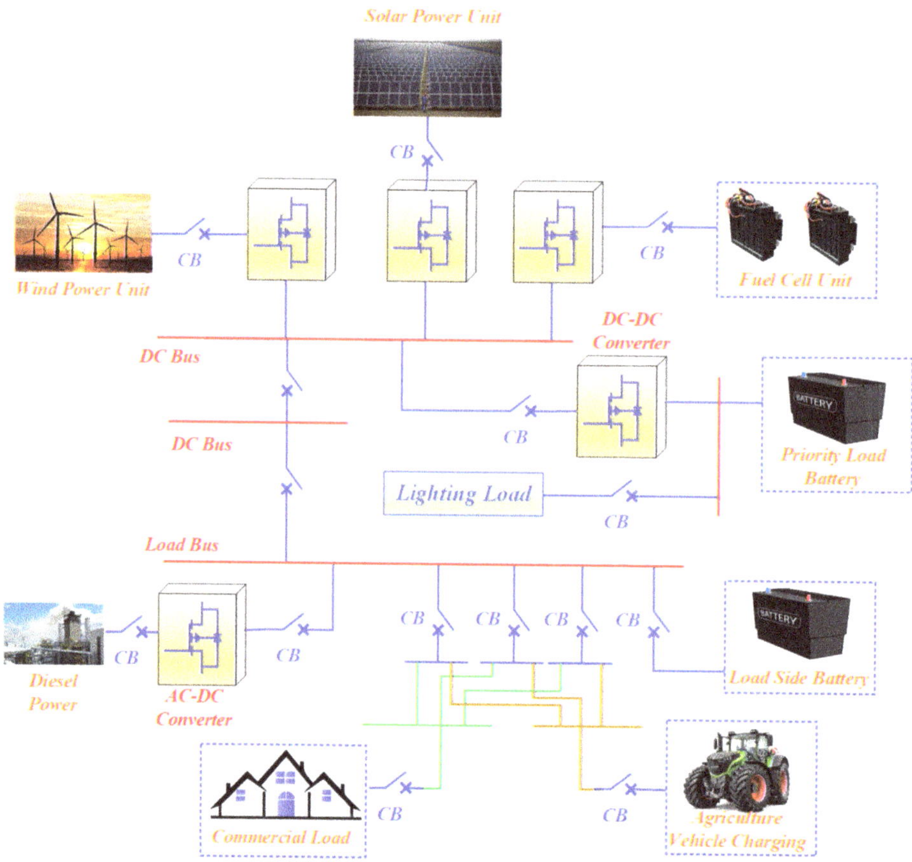

Figure 2. Proposed DC microgrid architecture.

To reduce voltage variation, a converter is used in all DG sources. The power produced from non-conventional energy sources areinconsistent to supply load continuously due to its intermittent. Therefore, batteries are utilized to ensure power supply to consumers without any interruption. Here, two batteries are used. One is interfaced to the load side, and the other is on the DC bus. The DC bus battery is only used for the priority load, and the load side battery is used for the commercial and agricultural vehicle loads. The diesel generator is also connected to the DC distribution network to deliver the load when available power in the DC microgrid is not sufficient. AC power generated by the diesel generator is converted to DC power by a converter.

3. Mathematical Modeling of ProposedDC Architecture

The proposed DC microgrid consists of four different sources (PV, wind, fuel cell, and battery). The PV power mainly depends upon solar irradiance and the ambient temperature. Wind power depends on wind velocity. PMSG is used to generate electrical power from the turbine. Wind power is interconnected to the DC microgrid by the AC/DC converter. Fuel cell power depends upon parameters such as distribution, flow rate, partial pressure of fuel cell gases, and the required number of stack cells. Mathematical modeling of each source is briefly discussed in this section.

3.1. Mathematical Modeling of Solar Power

The output power of the PV module is obtained by solar irradiance and with respect to PV module area. The output of the PV model is determined by [23]

$$P_{solar} = \eta_g i_r A, \tag{1}$$

where, η_g = generation efficiency, i_r = solar irradiation (W/m^2) and A = area (m^2), and the PV efficiency is determined by Equation (2).

$$\eta_{ce} = \eta_{ref}\eta_{ce}\left[1 - \beta\left(T_{cell} - T_{cellref}\right)\right], \tag{2}$$

where, η_{ce} = power conditioning efficiency, β = Temperature co-efficient C ((0.004–0.006)/C), η_{ref} = reference module efficiency, $T_{cellref}$ = reference cell temperature, and the temperature (T_C) is determined by Equation (3).

$$T_c = T_a + \left[\frac{NOCT - 20}{800}\right] G_t, \tag{3}$$

where, T_a = temperature in C, $NOCT$ = nominal operating cell temperature in C, G_t = solar irradiation in tilted module (W/m^2).

Total radiation in the solar cell considering normal and partial solar radiation is obtained by

$$T_I = I_D R_D + (I_b + I_d) R_r. \tag{4}$$

System Modeling

The PV or solar cell operation is similar to the operation of PN junction diode, which converts light energy into electricity through the photovoltaic effect [24]. The PV module is grouped based on the series and parallel connection of multiple PV cell [24,25]. The single PV cell is configured into a single diode representation as in Figure 3. In this model, solar irradiance is represented by a current source, and the other circuit parameters are diode current I_d, output current I, series resistance R_s, parallel resistance R_p, and output voltage V. The output current is calculated by

$$I = N_P\left[I_{ph} - I_{rs}\left(\frac{\exp q(V + IR_s)}{AKTN_s} - 1\right)\right], \tag{5}$$

$$I_{RS} = I_{rr} - \left[\frac{1}{T_K} - \frac{1}{T}\right], \tag{6}$$

where, N_P and N_S = number of cell connected in parallel and series, the K = Boltzmen constant, A = diode ideality factor, IR_S = reverse saturation current of cell at T, T_r = referred cell temperature, and I_{rr} = reverse saturation current at T_r

$$I_{ph} = \left[I_{scr} + K_i(T - T_r)\frac{S}{100}\right], \tag{7}$$

where, I_{scr} = short circuit current at a reference temperature of the cell, K_i = co-efficient of the short circuit temperature, S = solar irradiation in (W/m^2). In this model, the shunt resistance in parallel to the ideal shunt diode and the I-V characteristics are determined by the equation as follows:

$$I = I_{ph} - I_D, \tag{8}$$

$$I = I_{ph} - I_o\left[\exp\frac{q(V + IR_s)}{AKT} - 1\right] - \frac{V + R_s I}{R_{sh}}, \tag{9}$$

where, I_{ph} = Irradiance current (A), I_D = diode current (A), I_o = Inverse saturation current (A), R_S = series resistance (Ω), R_{sh} = shunt resistance (Ω), I = cell current (A), V = cell voltage, and the output current of the PV cell using single diode model is expressed as

$$I = I_{PV} - I_{D1} - \frac{V + IR_s}{R_{sh}}. \tag{10}$$

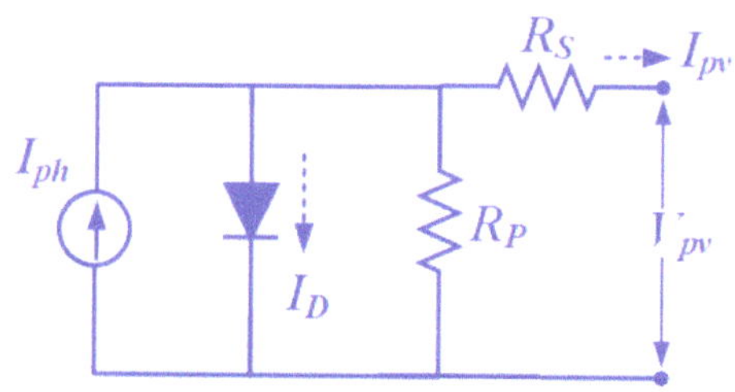

Figure 3. A single diodemodel of a photovoltaic (PV) cell.

The open circuit voltage and maximum power of the PV module is obtained by the simplified PV system modeling proposed by [26]. The voltage and power with the values of series resistance (R_S) is calculated by fill factor [27–29].

$$\left.\begin{array}{l} FF = FF_o\left[1 - \frac{R_S}{\frac{V_{OC}}{I_{SC}}}\right] \\ FF_o = \frac{V_{OC} - \ln(V_{OC} + 0.72)}{1 + V_{VOC}} \end{array}\right\} \tag{11}$$

$$\left.\begin{array}{c} P_{\max} = FF \times V_{OC} \times I_{OC} \\ P_{\max} = \frac{V_{OC} - \ln(V_{OC}+0.72)}{1+V_{OC}} \times \left(1 - \frac{I_{SC} \times R_S}{V_{OC}}\right) \times \frac{V_{OCO}}{1 + \beta \ln \frac{G_O}{G}} \times \left(\frac{T_O}{T}\right)^{\delta} \times I_{SO}\left(\frac{G}{G_O}\right)^{\alpha} \end{array}\right\} \tag{12}$$

where, FF = fill factor of the ideal PV module without resistive effects and V_{OC} = normalized value of the open circuit voltage to thermal voltage.

The power conversion in the PV system is obtained through the PV modules. The performance capability of the PV depends on the temperature and its characteristic curve (power & V, I curve) at standard test condition, which is shown in Figure 4. A single PV cell of any rating will not be able to generate the required power levels. Hence, several PV cells are interconnected through a series and parallel combinations that scale up to generate the required PV power. The voltage and current are obtained by scaling up of PV modules, which is expressed as

$$\left.\begin{array}{l} I_A = \frac{N_p}{I_M} \\ V_A = N_S \times V_M \\ P_A = FF \times V_A \times I_A \end{array}\right\} \tag{13}$$

where, I_A and V_A = PV array voltage and current, I_M and V_M = PV module voltage and current, and P_A and P_M = PV array power and module power.

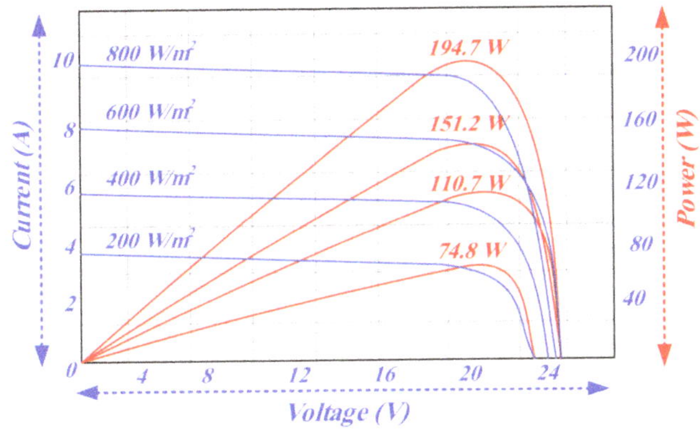

Figure 4. Solar cell characteristics curve (voltage vs. current and power).

3.2. Mathematical Modeling of Wind Power

The wind model consists of variation of wind velocity with gust and wind speed [30–36].

$$W_V = V_w + V_g + V_{wr}, \tag{14}$$

where, V_w = Base wind velocity, V_g = Gust wind velocity, and V_{wr} = ramp wind component.

The gust speed is calculated by

$$V_{wg} = \begin{Bmatrix} 0\ t < T_1 \\ C_2\left\{1 - \cos\pi\left[\frac{t-T_1}{T_2-T_1}\right]T_1 \le t \le T_2\right\} \\ 0\ t \ge T_2 \end{Bmatrix} \tag{15}$$

$$V_W = \begin{Bmatrix} 0\ s < T_3 \\ C_3\left\{\left[\frac{s-T_3}{T_4-T_3}\right]T_3 \le s \le T_4\right\} \\ 0\ s \ge T_4 \end{Bmatrix}, \tag{16}$$

where, C_2 = maximum value of the gust component, C_3 = maximum wind speed caused by the ramp, and T_3 and T_4 are the cut-in and cut-out times of the ramp, respectively.

Wind power is calculated by

$$P_W = \frac{dW_w}{dt} \tag{17}$$

Energy drawn by the wind turbine is

$$\left.\begin{array}{l} W_w = V_a \times \frac{1}{2}\rho(V_1{}^2 - V_3{}^2) \\ P_W = d\frac{V_a \frac{1}{2}\rho(V_1{}^2 - V_3{}^2)}{dt} \end{array}\right\} \tag{18}$$

where, W_w = energy drawn by wind turbine and ρ = Air density.

According to Betz, the maximum wind turbine power output is

$$P_M = \frac{16}{27}A_R\frac{3}{2}V^3 \tag{19}$$

Equation (19) is obtained by substituting the value for V_1, and V_3.

$$\begin{aligned} V_2 &= \tfrac{2}{3}V_1 \\ V_3 &= \tfrac{1}{3}V_1 \end{aligned} \tag{20}$$

The wind turbine model represents the output power captured by the turbine [33–36]. Figure 5 shows the characteristic curve for wind speed vs. power. The power in the wind (P_w) in an area is obtained by

$$P_W = \frac{1}{2}\rho A W_V{}^3 \tag{21}$$

$$P_M = P_W C_p \tag{22}$$

$$C_P = \frac{1}{2}[\delta - 0.22\beta^2 - 5.6)e^{-0.17\delta} \tag{23}$$

β = Pitch angle of the blade in degrees, δ = the tip speed ratio of the turbine, and C_p = Power coefficient.

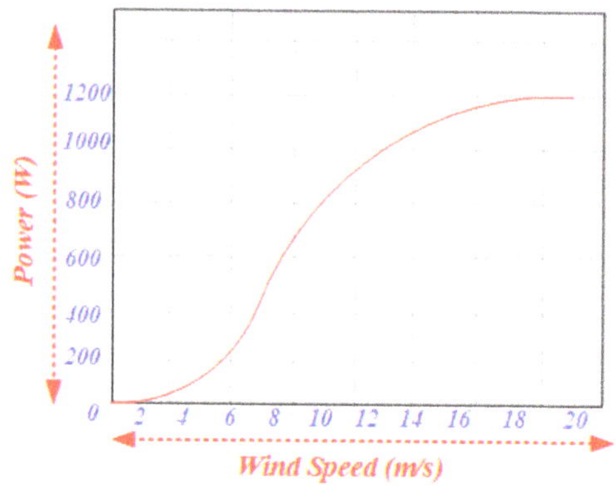

Figure 5. Characteristic curve of wind system (wind speed vs. power).

Wind generated power is expressed as:

$$p_G = V_G I_G \tag{24}$$

3.3. Mathematical Modeling of Fuel Cell Power

A different assumption [36] is made, which is described below:

- Idealized modeling;
- Uniform circulated gases;
- Constant pressure in the flow channel;
- Cell parameters are represented together to form stack parameters;
- The output voltage of the single fuel cell can be represented as

$$\left.\begin{aligned} PH_2 &= \tfrac{-1}{tH_2}\left[PH_2 + \tfrac{1}{KH_2}\left(q_{th}{}^{in} - 2K_r i_{FC}\right)\right] \\ PO_2 &= \tfrac{-1}{tO_2}\left[PO_2 + \tfrac{1}{KO_2}\left(q_{th}{}^{in} - 2K_V i_{FC}\right)\right] \\ PH_2O &= \tfrac{-1}{tH_2O}\left[PH_2O + \tfrac{1}{KH_2O}K_r i_{FC}\right] \end{aligned}\right\} \tag{25}$$

Here, KH_2 = valve meter constant for hydrogen and KO_2 = valve meter constant for oxygen (2.52×10^{-3} kmol/s atm). K_r = constant defined by the rate of reactant hydrogen and fuel cell current. The reactant utilization factor U is defined as follows:

$$U_{\Gamma} = \frac{qH_2{}^{in} - qH_2{}^{out}}{qH_2{}^{in}} \tag{26}$$

$$\begin{aligned} V_{act} &= [E_1 + E_2T + E_3Tln(co_2) + E_4ln(i_{FC})] \\ co_2 &= \frac{po_2}{5.08 \times 10^6 e\left(\frac{498}{T}\right)} \end{aligned} \tag{27}$$

where, E_1, E_2, E_3, E_4 is the cell parameter coefficients, CO_2 = concentration of oxygen. Figure 6 represents the equivalent circuit of the fuel cell. It consists of cell voltage, actual resistance, concentration resistance, and ohmic resistance [37,38].

$$V_{ohm} = i_{FC}(R_m + R_{ohm}) \tag{28}$$

$$R_m = \frac{\rho_m l}{A} \tag{29}$$

where, ρ_m = specific resistivity of the membrane for electron flow (Ω cm). A = active cell area (cm^2) and l = thickness of the membrane.

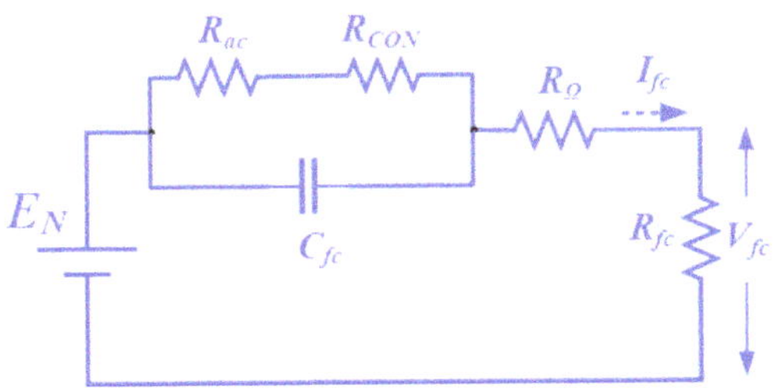

Figure 6. Equivalent circuit of the fuel cell.

The concentration loss is due to the reactive excess concentration near the catalyst surface.

$$V_{Con} = \frac{-R_{con}T}{2F} \ln\left(1 - \frac{j}{j_{\max}}\right) \tag{30}$$

The fuel cell current can be determined as

$$i = i_0 A\left(e^{\frac{\alpha nf}{P_{fC}}} V_{act} - e^{\frac{(1-\alpha)nF}{RTFC} V_{ac}}\right) \tag{31}$$

I_o = exchange current density (A/m^2), A = catalyst layer surface (m^2), and i = fuel cell current. The Figure 7 shows the single fuel cell characteristics for stack current vs. cell voltage and power. The power of the fuel cell can be obtained from

$$P_{fuel} = V_{out} i_C. \tag{32}$$

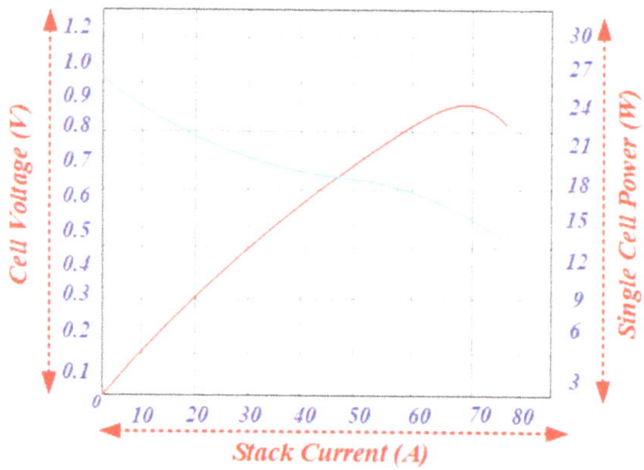

Figure 7. Single fuel cell characteristics curve (stack current vs. cell voltage and power).

3.4. Mathematical Modeling of the Battery

The mathematical models of the battery focus mainly on V, I parameters. The current is determined by a change in the terminal voltage of the battery [36]. The transfer of electrons from one electrode to another leads to the generation of current. The open circuit voltage at the battery is determined from the potential difference between the positive and negative electrodes [37–40]. The charging/discharging of battery is expressed as

$$V_{discharge} = E_o - V_{op}{}^{+} - V_{op} - IR_{pol} \tag{33}$$

$$V_{charge} = E_o + V_{op}{}^{+} + V_{op} + IR_{pol} \tag{34}$$

$$V_{battery} = E_o - K\left[\frac{Q}{Q - it}\right]i - R_o i \tag{35}$$

$$V_{battery} = E_o - \left(\frac{K}{Soc}\right)i - R_o i \tag{36}$$

$$V_{discharge} = E_o - Kd_r\frac{Q}{Q - it}i_1 - R_o i - Kd_v\frac{Q}{Q - it}it + e(t) \tag{37}$$

$$V_{charge} = E_o - Kc_r\frac{Q}{it + \lambda Q}i_1 - R_o i - Kc_v\frac{Q}{Q - it}it + e(t) \tag{38}$$

$$e(t) = Bi[(e(t) + Au(t)] \tag{39}$$

$$V_{dicharge} = E_o - K_{dr}\frac{1}{Soc}i - R_o i - K_V(\frac{1}{Soc} - 1) + e(t) \tag{40}$$

Equation (35) can be rewritten using state of charge (SoC) due to the polarization ohmic voltage. Equations (37) and (38) are modified by the shepherd relation model. E_o = open circuit voltage of a battery (V), K = polarization coefficient (Ω), Q = battery capacity (A/h), and R = internal resistance. Some of the limitations associated with Equations (37) and (38) are (i) ageing of battery and self-discharge, (ii) the battery capacity does not depend upon the amplitude of the current, and (iii) the temperature coefficient is not considered [35]. These limitations can be overcome by considering the factors affecting the lifetime of the battery. The SoC condition is analyzed at every instant of time and is calculated with threshold capacity using

$$Soc = Soc_{in} - \int_{0}^{t} (i - \max(i_g, i_d))\frac{d\tau}{Q}. \quad (41)$$

The net power of the DC microgrid architecture is calculated by the summing of all the power of the energy sources.

$$P_{net} = P_{PV} + P_{Wind} + P_{fuelcell} + P_{diesel} \quad (42)$$

4. Energy Management Strategy

The proposed DC microgrid architecture has three renewable energy sources with the storage device and diesel generator to supply the load continuously. In villages, agriculture is the main occupation of people. Thus, the microgrid is designed to meet the commercial DC and agricultural vehicle load. The lighting of the village will be considered as a priority load in the system. The net generated power and load power P_{net} and P_L is given by Equations (43) and (44).

$$P_{net} = P_{PV} + P_{Wind} + P_{fuelcell} \quad (43)$$

$$P_L = P_{DL} + P_{AGL} + P_{PL} \quad (44)$$

where,

P_{PV} = power generated by PV (kW);
P_{wind} = power generated by Wind (kW);
$P_{fuelcell}$ = power generated by Fuel cell (kW);
P_L = Domestic load (kW);
P_{AGL} = Agriculture vehicle load (kW);
P_{PL} = Priority load (kW).

The priority battery is used to supply the lighting load.It will charge during the day and discharge during the night. The generated power will supply the loads by the three cases with the help of the battery and diesel generator. The process is represented as a flow chart in Figure 8. First, the load demand and generated by various sources will be measured based on the condition as follows

- Case 1: Generated power equal to total load.

In this condition, the power generated by wind, PV, and fuel cell is equal to the total load, hence the load will be supplied by the generation without any interruption.

- Case 2: Generated power higher than the load.

In this case, the generated power is higher than the load, hence the renewable power is fully supplied to the required load. The excess power from the generation is charged to the battery. The proposed architecture has two batteries: one is the priority load battery, and the other is the domestic and agricultural load side battery.While the load is supplied, at the same time, the priority battery SoC will be checked. Thus, the condition of priority the load battery and domestic load battery is discussed in Case 3 and Case 4.

- Case 3: Priority load battery charging.

In this case, the SoC of the priority load battery is measured. If the priority is minimum, the battery is charged until it reaches the rate power on that interval. Once it reaches the rated power, the priority battery will not charge, and the excess power generated by the energy source will charge the commercial battery.

$$SoC_{min} < SoC_{priority} < SoC_{max}$$

$$P_{power} > P_{rated} = charging$$

$$P_{power} < P_{rated} = discharging$$

$$P_{priority} = \left\{ \begin{array}{ll} 7\text{ a.m.} < charging < 6\text{ p.m.} & \text{rated (kW/h)} \\ 6\text{ p.m.} < discharging < 7\text{ a.m.} & \text{rated (kW/h)} \end{array} \right\}$$

where, P_{power} = priority load power P_{rated} = rated (kW/h) to be charged on the condition of $P_{priority}$ interval, charging on daytime P_{rated} = rated (kW/h) to be discharged on the condition of $P_{priority}$ interval discharging on night time.

- Case 4: Domestic and agricultural charging.

In this case, the *SoC* of the commercial load battery is measured, if the $SoC_{Combattery}$ is minimized then the battery is charged till the *SoC* of the battery reaches the maximum value. Once it is fully charged excess power generated will be reduced by controlling the output of the fuel cell.

$$SoC_{min} < SoC_{Combattery} < SoC_{\max}$$

$$SoC_{Combattery} > SoC_{min} = charging$$

$$SoC_{Combattery} > SoC_{\max} = discharging$$

- Case 5: $P_G < P_L$.

The generated power will be less than the required load, in this case. This case will be treated with caution to supply the load with the help of the battery by cases 6 and 7. Whatever the load profile may be, the priority load will be supplied by generation, which is the major consideration in this case. The difference in power from the generation and load will be calculated, and then whether the available generation is enough to meet the priority load will be checked. When the condition is satisfied, the priority load will be met by the available generation.

$$P_{min} < P_{priority} < P_{max}$$

$$P_{net} = P_{min} > P_{rated} \text{ (i.e., checks the condition for the case 6)}$$

$$P_{net} = P_{max} < P_{rated} \text{ Supply the load}$$

- Case 6: Checking the condition of commercial battery $P_{Combattery}$.

In this case, the commercial load battery power is measured.The generated power and the commercial load battery are checked to see if they are able to supply the load. If the power is sufficient, the demand will be supplied until the *SoC* of the commercial load battery reaches a minimum level.

$$SoC_{min} < SoC_{Combattery} < SoC_{max} SoC \text{ for the commercial battery}$$

$$SoC_{Combattery} < SoC_{max} \text{ Battery is discharging}$$

$$P_L = P_{Combattery} + P_G$$

- Case 7: Checking the total demand with respect to the generator.

If $P_L = P_{Combattery} + P_G$, then the demand will be supplied through the diesel generator, and the system will continuously check the generative power and the priority load. Once the generation power is enough to supply the load, the diesel generator is cut off from the grid.This process is continued to give uninterrupted power to consumers in remote villages.

$$P_L = P_{combattery} + P_G + P_{Diesel} \tag{45}$$

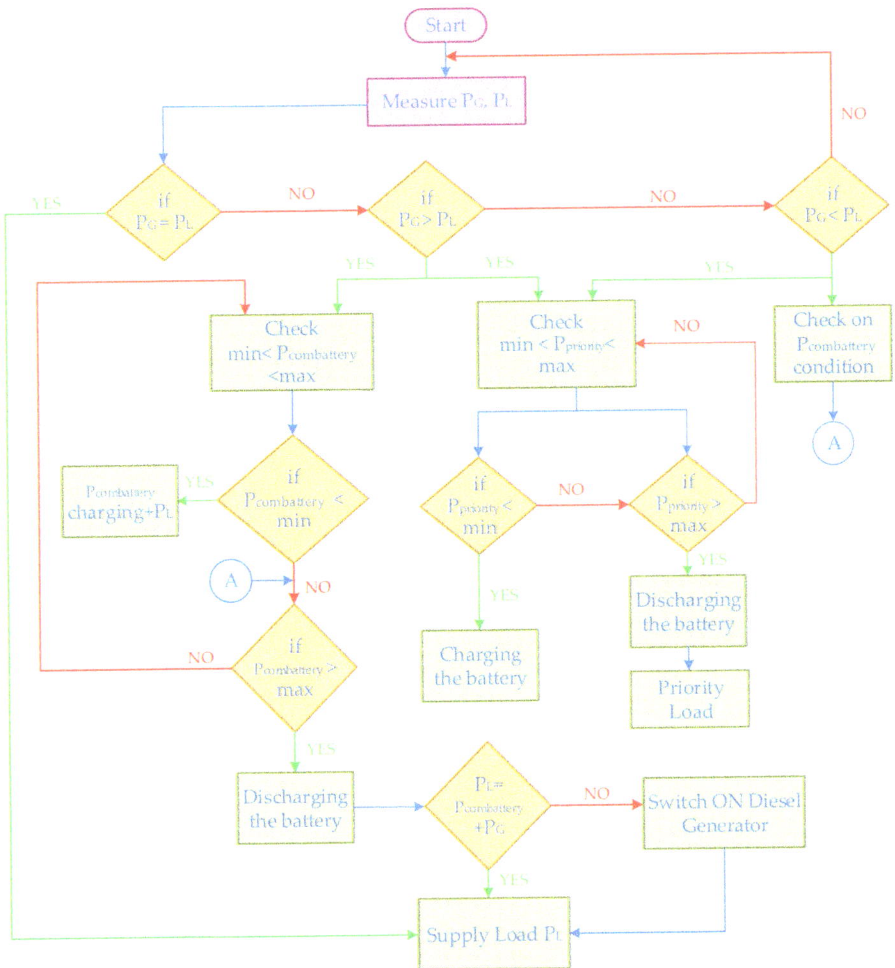

Figure 8. Flowchart for proposed energy management strategy (EMS) of the DC microgrid.

5. Simulation Study and Results

5.1. Simulation Results

The proposed EMS is simulated for the DC microgrid using MATLAB/Simulink. The simulation parameters are shown in Table 1. The control parameter for the EMS is PV, wind, fuel cell, battery, diesel generator, and load power. The load is subdivided into three categories: priority, commercial, and agricultural vehicle load.

Table 1. Simulation parameters.

Description	Specification
Wind generator	5 kW, 220 V
PV	5 kW, 220 V
Fuel Cell	5 kW, 220 V
DC bus voltage	220 V
DC-DC converter	220 V
Dynamic Load	5 kW
Motor load	2.5 kW to 5 kW
Battery	220 V battery/150 Ah

The priority load consists of the basic requirement of load that has to be supplied continuously (street lighting of village community), whereas the commercial load and agricultural vehicle load vary according to the requirements of the consumer. Based on the available power generated by all sources with respect to load, the EMS is categorized into the following three cases:

1. Generation equal to load ($P_G = P_L$);
2. Generation greater than load ($P_G > P_L$);
3. Generation less than load ($P_G < P_L$).

5.2. Generation Equal to Load ($P_G = P_L$)

The characteristic curve shown in Figure 9 illustrates that the available power generation is equal to the load condition that is plotted between times versus power. From this condition, the renewable power energy source power is completely utilized to meet the demand requirements.In this condition, the battery will not charge/discharge because the power is deficient. It will not be necessary to use the diesel generator in this case. This case only occurs during a few hours a day, which is shown in Table 2.

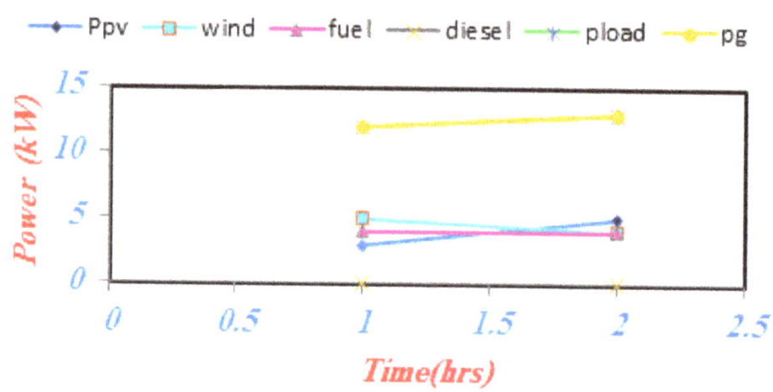

Figure 9. Generation is equal to the load ($P_G = P_L$) (power vs. time).

5.3. Generation Greater Than Load (PG > PL)

The Figure 10 shows that the available power generation is greater than load. When $P_G > P_L$, the available power is supplied to the consumers and excess power is stored in batterries. However, before storing energy in the batteries, the EMS checks the *SoC* of both the batteries. First, the priority load battery *SoC* will be checked and be based on the requirement in a particular interval. It will be charged, and next, the load-side battery *SoC* will be measured based on the level of *SoC*. When the *SoC* is minimized, the power is not delivered to the load; rather, it is stored in the batteries. In certain cases, the power will be extremely high and the load is quite low.Then, the battery is charged completely. There will be surplus power.Thus, the generation of power is cutoff.

Table 2. Simulated result values ("kW") of the DC microgrid with EMS.

Time	Power Generated by Renewable Energy Source			Diesel Power	Total Generated Power	Commercial Battery		Priority Load Battery		Priority Load	Domestic Loads		Agricultural Loads		Total Loads
	Ppv (kW)	Wind (kW)	Fuel (kW)	Diesel (kW)	P_g (kW)	Charge (kW/h)	Discharge (kW/h)	Charge (kW/h)	Discharge (kW/h)	Load (kW)	P_{d1} (kW)	P_{d2} (kW)	P_{ag1} (kW)	P_{ag2} (kW)	P_L (kW)
6 a.m.–7 a.m.	1	3	5	4	9	0	0	−2	0	2	3	2	4	4	15
7 a.m.–8 a.m.	3	4	4	0	11	−1	0	−2	0	2	2	2	2	2	10
8 a.m.–9 a.m.	4	5	4	0	13	−1	0	−2	0	2	1	2	4	3	12
9 a.m.–10 a.m.	3	5	4	0	12	0	0	−2	0	2	4	0	3	3	12
10 a.m.–12 p.m.	5	4	5	0	14	−2	0	−2	0	2	2	2	3	3	12
12 p.m.–2 p.m.	5	5	5	0	15	−7	0	−2	0	2	2	1	3	0	8
2 p.m.–3 p.m.	4	2	3	0	9	0	3	−2	0	1	2	3	3	3	12
3 p.m.–4 p.m.	5	4	4	0	13	0	0	−1	0	1	4	1	4	3	13
4 p.m.–6 p.m.	3	5	3	0	11	−9	0	1	3	−3	1	1	1	1	4
6 p.m.–8 p.m.	1	2	4	0	7	0	4	2	3	−3	4	3	2	2	11
8 p.m.–12 a.m.	0	4	5	0	9	0	1	2	3	−3	1	3	2	4	10
12 a.m.–4 a.m.	0	4	5	0	9	−4	0	2	3	−3	3	2	0	0	5
4 a.m.–6 a.m.	0	0	2	0	2	0	9	2	3	−3	2	2	5	2	11

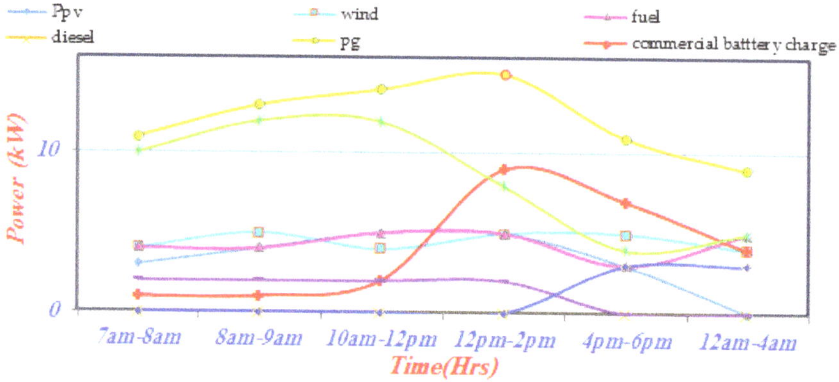

Figure 10. Generation greater than load ($P_G > P_L$) (power vs. time).

5.4. Generation Greater Than Load ($P_G < P_L$)

Figure 11 illustrates the condition where the available power generated is less than the load. During ($P_G < P_L$), the EMS checks whether the power generated is able to supply at least the priority load or not. If the generated power is sufficient to supply the demand of priority load, the EMSthen checks the condition of the battery. The priority load will be supplied by the generators. For supplying the commercial and agricultural vehicle load, the battery will supply power when the required power is already stored in the battery. If the battery power and generated power is not able to supply the load, the diesel generator will then supply the load power. The performance of the proposed EMS is analyzed for a 24 h period, which is shown in Table 2. From that table, it is observed that the load requirement of the rural village community is supplied by the penetration of various renewable energy sources rather than the diesel generator. The diesel consumption of the DC microgrid is only 4 kW for one hour in a whole day compare to all other EMS.

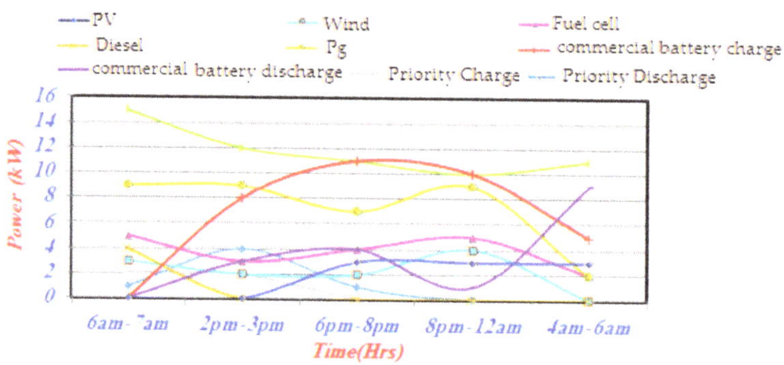

Figure 11. Generation greater than load ($P_G > P_L$).

The complete performance of the proposed system for 24 h period of the EMS of the DC microgrid is summarized in Table 3, and Figure 12 describes the PV, wind, fuel cell, and diesel power and their operating condition with respect to load in all cases. The laboratory-scale DC microgrid has a capacity of 750 W, and with each renewable energy source, the capacity is 250 W.The performance of all the sources depends on the climatic condition. The power upstream and downstream of the DC microgrid depends on the maximum and minimum voltage, current rating, with respect to the standard test condition (STC) of individual sources stated by the manufacturer. Based on that, the DC microgrid is designed, and their specification is described in Table 4.

Table 3. Summary of the EMS.

	P_G	P_L	$P_{Priority}$	P_{Com}	P_{Diesel}	Cases	Remarks
1	$P_G = P_L$	√	√	√	×	1	The power from renewable source is enough to supply the load without storage unit and diesel generator.
2	$P_G > P_L$	√	√	√	×	2,3,4	The load will be supplied and additionally the batteries will charge with the surplus power
3	$P_G < P_L$	√	√	√	√	5,6,7	The load is supplied by the combination of diesel generator, available power, and battery.

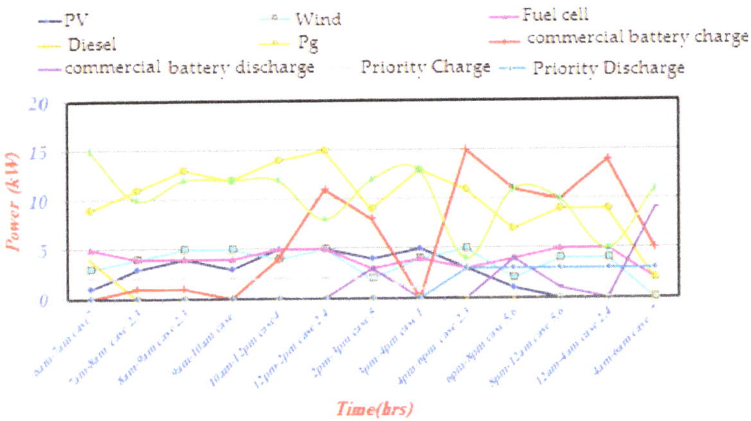

Figure 12. Overall performance of the system with generated power.

Table 4. Hardware specifications of the proposed DC microgrid.

PV		Wind		Fuel Cell	
Centsys Solar 250 W		SIKCO Wind 1000		Horizon 500 W PEM Fuel Cell	
PV Modules	Specification	Wind	Specification	Fuel Cell	Specification
Maximum capacity	250 W	Type of Turbine	Horizontal Axis Downwind Turbine	Rated capacity	500 W
Tolerance	±3%	voltage	12 V DC	Rated voltage	14.4 V
Open circuit voltage	37.8 V	Rated Wind Speed	5 m/s	Valve Voltage	12 V
Short circuit current	7.94 A	Rated Power	300 W	Blower range	12 V
Module efficiency	15.3%	Rated rpm	300	Reactants	Hydrogen and Air
Solar cell efficiency	17.2%	Cut-in wind speed	2 m/s	Ambient Temperature	5–30 °C (41–86 °F)
Maximum voltage (V_m)	31.5 V	Cut-out wind speed	15 m/s	Max Stack Temperature	65 °C (149 °F)
Maximum current (I_m)	7.94 A	Blade length	600 mm	Gas Pressure	0.45–0.55 Bar
Nominal Temperature	42 °C (±2 °C)	blades	6	Stack Size	268 mm × 130 mm × 122.5 mm (10.5″ × 5.1″ × 4.8″)
Dimensions	1650 mm × 992 m × 40 mm	Noise Level	<20 dB	Efficiency of System	40% at 14.4 V

6. Experimental Analysis

The performance of the proposed EMS is verified by the laboratory-scale DC microgrid. The output power of the PV and fuel cell are DC, but for output power is AC.Thus, the AC power is converted to DC with the help of converters. All three sources are integrated to a common DC bus, and a charge controller is used to maintain the voltage level. Two batteries are connected in parallel to supply the demand. One is the priority load battery, and the other is the commercial load battery. The charging and discharging modes of the batteries are based on the availability of power in the DC microgrid. The hardware description for the DC microgrid is presented in Table 5.

Table 5. Hardware description of components used.

Parameters	Specification
DC bus voltage	24 V
Capacityof wind generator	200 W
Capacity of PV panel	200 W
Capacity of fuel cell power	100 W
Battery type	Tall tubular C10
Battery capacity	14 Ah/12 V
DC-DC converter	24 V/220 V
Lamp loads	500 W
Load bus	220 V
Diesel generator	500 W
Maximum current	3 A

The common DC bus voltage of the laboratory scale system is 24 V, and the 24 V/220 V converter is used to maintain the voltage level at load bus. The EMS is tested under the following cases:

1. $P_G > P_L$
2. $P_G = P_L$
3. $P_G < P_L$

The EMS continuously monitors the load power with respect to the renewable power generation. Based on its performance, the DC microgrid will supply the load. Two types of loads are considered to validate the performance: one is the priority load, and the other is the commercial load.

The priority load is the lighting load that will charge during the day and discharge at night. The commercial battery is charged based on the availability of RES with respect to the load. The experimental results are shown in Table 6.

(1) Case 1: $P_G > P_L$

During 7 a.m.–8 a.m., the total power generated from the REG is 370 W. At that instant, the demand is 250 and the load current is 1.2 Amps. The excess power (120 W) is stored in the commercial load battery. The voltage of the DC bus and the load current waveform is shown in Figures 13 and 14.

(2) Case 2: $P_G = P_L$

When the load is raised at 9 a.m.–10 a.m. to generate power, the load power is 440 W and the load current is 2 A, and the generated power continuously supplies the load without any interruption. At that interval, the commercial battery is in standby mode (no charging and discharging). The load current and priority load battery charging waveform are shown in Figures 15 and 16.

(3) Case 3: $P_G < P_L$

Normally, PV does not supply power from 6 p.m.–6 a.m. due to the sun irradiation portfolio. During this time, the rest of power sources (wind and fuel cell) combine with battery to supply the load. At 8 p.m.–12 a.m., the load is considered to be 900 W and the available generation is 600 W.

In this case, the microgrid power demand is satisfied through a combination of battery and available energy sources (wind and fuel cell). The load current, commercial load battery, and priority load battery charging waveform are shown in Figures 17 and 18. If the total generation on the microgrid is low when compared to the load, the diesel generator will switch on to supply the load until the renewable power becomes active. From all the three cases, this system will be reliable and supply the load without any interruption. The main advantage of this DC micro grid is that it will supply the load to the consumer with the depth penetration of renewable energy sources rather than the diesel generator. It is concluded that the prototype of the 500 W microgrid is tested and validated with different conditions. The results satisfy the EMS for both the simulation and the hardware setup. The DC microgrid architecture is scalable for 15 kW to satisfy the load demand in rural communities.

Table 6. Experimental results at different time periods.

Time	PV Power (W)	Wind Power (W)	Fuel Cell Power (W)	Generated Power (W)		Loads (W)						Batteries			
						Priority Load		Commercial Loads		Net Loads		Priority Load Battery		Commercial Load Battery	
				P(W)	I(A)	P(W)	I(A)	P(W)	I(A)	P(W)	I(A)	P(W)	I(A)	P(W)	I(A)
7 a.m.–8 a.m.	100	170	100	370	1.68	100	0.5	150	0.7	250	1.2	−100	−0.5	−120	−0.7
9 a.m.–10 a.m.	200	140	100	440	2	100	0.5	340	1.5	440	2.0	−100	−0.5	0	0
8 p.m.–12 a.m.	0	150	100	250	1.13	100	0.5	250	1.1	350	1.6	100	0.5	100	0.45

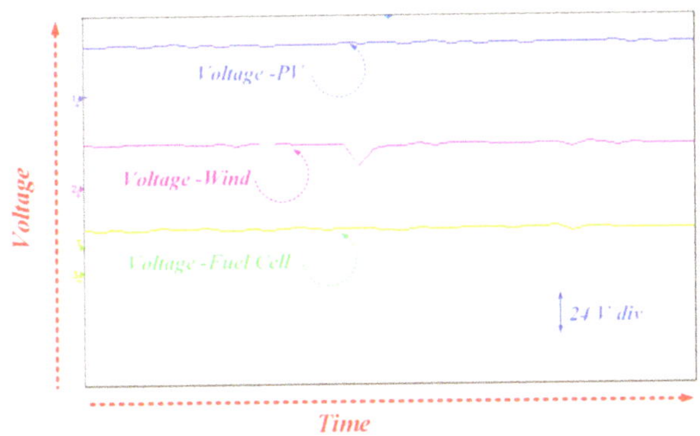

Figure 13. Voltage across each renewable energy source.

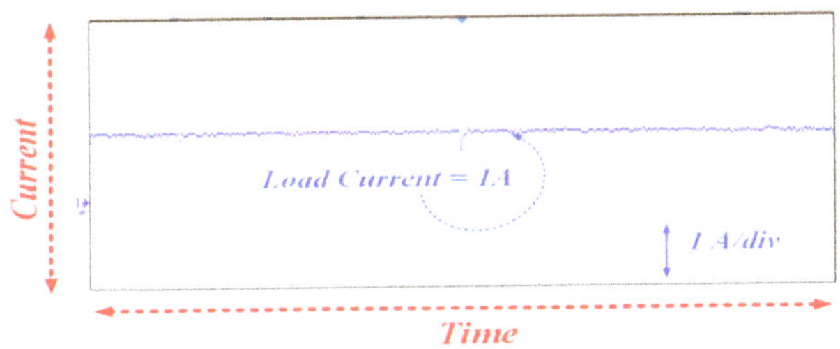

Figure 14. Load current during 7 a.m.–8 a.m.

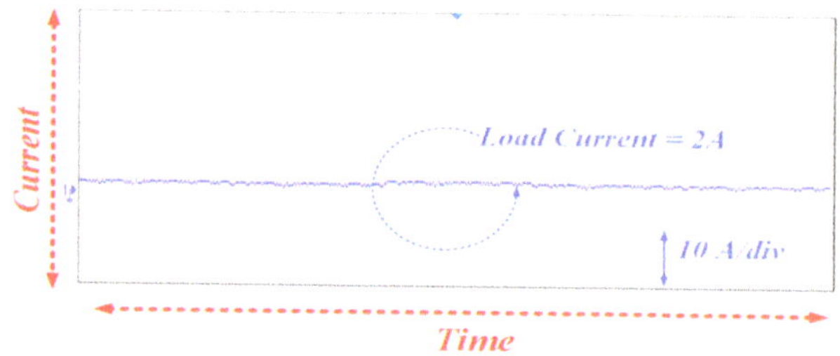

Figure 15. Load current during 9 a.m.–10 a.m.

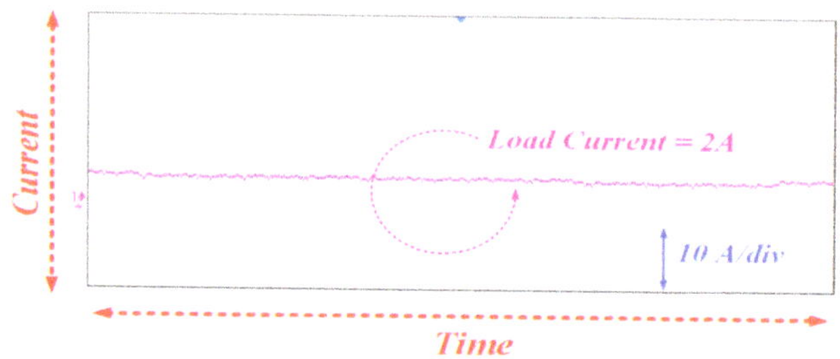

Figure 16. Load current during 8 p.m.–12 a.m.

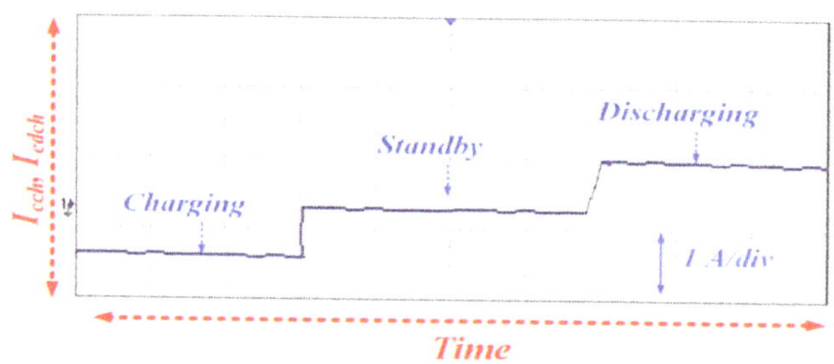

Figure 17. Commercial battery charging/discharging profile.

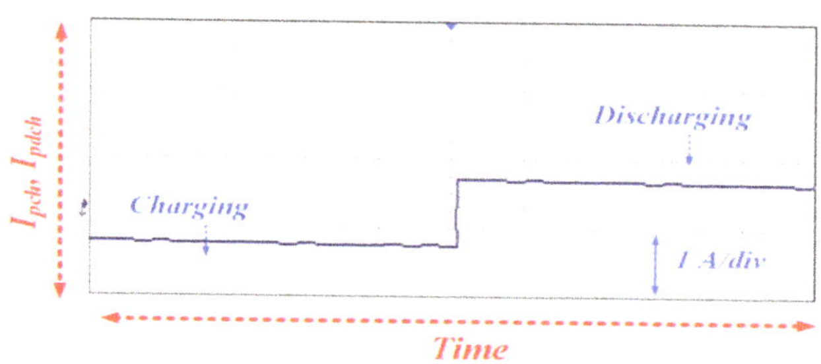

Figure 18. Priority load battery charging/discharging profile.

7. Conclusions

In this paper, an EMS is proposed for different renewable sources fedtoDC microgrid for remote village communities with capricious load conditions. The Proposed DC microgrid handles the load power balancing for DG sources based on EMS. The proposed system is applicable for electrifying rural communities with maximum penetration of renewable energy sources and storage systems. Furthermore, this EMS is able to handle the load–power balancing for all the capricious cases ($P_G = P_L$, $P_G > P_L$ and $P_G < P_L$) and providesa continuously supply to rural communities (described in Table 2 for all capricious cases). The load–power balancing is performed based on two battery sources: one is the priority battery (which handles the priority load), and the other is the commercial battery (which isapplicable to common DC bus). This battery is able to handle the load demand for various capricious cases, which are described in Table 2. Hence, the EMS utilizes the maximum power from the renewable sources and reduces the consumption of non-conventional energy sources (a diesel generator). The simulation and experimental studies of the DC microgrid with the proposed EMS clearly indicates that the power dissipates to the consumer through maximum renewable energy penetration and batteries throughout the day without any divergence in the system. Thus, the proposed EMS is verified through a laboratory-scale real time DC microgrid experimental setup and confirms its merits.

Author Contributions: Maheswaran Gunasekaran and Hidayathullah Mohamed Ismail has developed the proposed research concept, and they both are involved to study the execution and implementation with numerical software by collecting information from the real environment and developed the simulation model for the same. Maheswaran Gunasekaran has implemented the proposed concept in laboratory prototype environment with Bharatiraja Chokkalingam. Sanjeevikumar Padmanaban, and Lucian Mihet-Popa shared thier expertise and validation examinations to confirm the concept theoretically with the obtained numerical results for its validation of the proposal. All authors Maheswaran Gunasekaran, Hidayathullah Mohamed, Bharatiraja Chokkalingam, Sanjeevikumar Padmanaban, Lucian Mihet-Popa, involved to frame the final version of the manuscript as a full research article. Moreover, all authors involved in validating and to make the article error free technical outcome for the set investigation work.

Conflicts of Interest: The authors declare no conflict of interest.

References

1. Bloomberg New Energy Finance. *The Future of Energy 2012 Results Book*; Bloomberg New Energy Finance: New York, NY, USA, 2012.
2. International Energy Agency World Energy Investment. Available online: https://www.iea.org/media/publications/investment/WEI2017Launch_forWEB.pdf (accessed on 10 April 2017).
3. Anand, S.; Fernandes, B.G. Steady state performance analysis for load sharing in DC distributed generation system. In Proceedings of the 2011 10th International Conference on Environment and Electrical Engineering (EEEIC), Rome, Italy, 8–11 May 2011; pp. 1–4.
4. Gao, L.; Liu, Y.; Ren, H.; Guerrero, J.M. A DC Microgrid Coordinated Control Strategy Based on Integrator Current-Sharing. *Energies* **2017**, *10*, 1116. [CrossRef]
5. Patterson, B.T. Dc, come home: Dc microgrids and the birth of the 'enernet'. *IEEE Power Energy Mag.* **2012**, *10*, 60–69. [CrossRef]
6. Ali, A.; Padmanaban, S.; Twala, B.; Marwala, T. Electric Power Grids Distribution Generation System for Optimal Location and Sizing—A Case Study Investigation by Various Optimization Algorithms. *Energies* **2017**, *10*, 960. [CrossRef]
7. Liu, X.; Wang, P.; Loh, P.C. A hybrid AC/DC microgrid and its coordination control. *IEEE Trans. Smart Grid* **2011**, *2*, 278–286. [CrossRef]
8. Sechilariu, M.; Wang, B.C.; Locment, F. Supervision control for optimal energy cost management in DC microgrid: Design and simulation. *Int. J. Electr. Power Energy Syst.* **2014**, *58*, 140–149. [CrossRef]
9. Hossain, E.; Perez, R.; Padmanaban, S.; Mihet-Popa, L.; Blaabjerg, F.; Ramachandaramurthy, V.K. Sliding Mode Controller and Lyapunov Redesign Controller to Improve Microgrid Stability: A Comparative Analysis with CPL Power Variation. *Energies* **2017**, *10*, 1959. [CrossRef]

10. Subramani, G.; Ramachandaramurthy, V.K.; Padmanaban, S.; Mihet-Popa, L.; Blaabjerg, F.; Guerrero, J.M. Grid-Tied Photovoltaic and Battery Storage Systems with Malaysian Electricity Tariff—A Review on Maximum Demand Shaving. *Energies* **2017**, *10*, 1884. [CrossRef]
11. Tan, K.M.; Ramachandaramurthy, V.K.; Yong, J.Y.; Padmanaban, S.; Mihet-Popa, L.; Blaabjerg, F. Minimization of Load Variance in Power Grids—Investigation on Optimal Vehicle-to-Grid Scheduling. *Energies* **2017**, *10*, 1880. [CrossRef]
12. Sen, R.; Bhattacharyya, S.C. Off-grid electricity generation with renewable energy technologies in India: An application of HOMER. *Renew. Energy* **2014**, *62*, 388–398. [CrossRef]
13. AL-Nussairi, M.K.; Bayindir, R.; Padmanaban, S.; Mihet-Popa, L.; Siano, P. Constant Power Loads (CPL) with Microgrids: Problem Definition, Stability Analysis and Compensation Techniques. *Energies* **2017**, *10*, 1656. [CrossRef]
14. Ganesan, S.; Padmanaban, S.; Varadarajan, R.; Subramaniam, U.; Mihet-Popa, L. Study and Analysis of an Intelligent Microgrid Energy Management Solution with Distributed Energy Sources. *Energies* **2017**, *10*, 1419. [CrossRef]
15. Song, M.; Chen, K.; Zhang, X.; Wang, J. Optimization of wind turbine micro-siting for reducing the sensitivity of power generation to wind direction. *Renew. Energy* **2016**, *85*, 57–65. [CrossRef]
16. Dixon, C.; Reynolds, S.; Rodley, D. Micro/small wind turbine power control for electrolysis applications. *Renew. Energy* **2016**, *87*, 182–192. [CrossRef]
17. Macedo, W.N.; Monteiro, L.G.; Corgozinho, I.M.; Macêdo, E.N.; Rendeiro, G.; Braga, W.; Bacha, L. Biomass based microturbine system for electricity generation for isolated communities in amazon region. *Renew. Energy* **2016**, *91*, 323–333. [CrossRef]
18. Chokkalingam, B.; Padmanaban, S.; Siano, P.; Krishnamoorthy, R.; Selvaraj, R. Real-Time Forecasting of EV Charging Station Scheduling for Smart Energy Systems. *Energies* **2017**, *10*, 377. [CrossRef]
19. Bonfiglio, A.; Brignone, M.; Invernizzi, M.; Labella, A.; Mestriner, D.; Procopio, R. A Simplified Microgrid Model for the Validation of Islanded Control Logics. *Energies* **2017**, *10*, 1141. [CrossRef]
20. Long, B.; Jeong, T.W.; Deuk Lee, J.; Jung, Y.C.; Chong, K.T. Energy management of a hybrid AC–DC micro-grid based on a battery testing system. *Energies* **2015**, *8*, 1181–1194. [CrossRef]
21. Martin-Martínez, F.; Sánchez-Miralles, A.; Rivier, M. A literature review of Microgrids: A functional layer based classification. *Renew. Sustain. Energy Rev.* **2016**, *62*, 1133–1153. [CrossRef]
22. Shahzad, M.K.; Zahid, A.; Ur Rashid, T.; Rehan, M.A.; Ali, M.; Ahmad, M. Techno-economic feasibility analysis of a solar-biomass off grid system for the electrification of remote rural areas in Pakistan using HOMER software. *Renew. Energy* **2017**, *106*, 264–273. [CrossRef]
23. Olatomiwa, L.; Mekhilef, S.; Ismail, M.S.; Moghavvemi, M. Energy management strategies in hybrid renewable energy systems: A review. *Renew. Sustain. Energy Rev.* **2016**, *62*, 821–835. [CrossRef]
24. Mandelli, S.; Barbieri, J.; Mereu, R.; Colombo, E. Off-grid systems for rural electrification in developing countries: Definitions, classification and a comprehensive literature review. *Renew. Sustain. Energy Rev.* **2016**, *58*, 1621–1646. [CrossRef]
25. Hosseini, S.A.; Abyaneh, H.A.; Sadeghi, S.H.H.; Razavi, F.; Nasiri, A. An overview of microgrid protection methods and the factors involved. *Renew. Sustain. Energy Rev.* **2016**, *64*, 174–186. [CrossRef]
26. Jing, W.; Lai, C.H.; Wong, S.H.W.; Wong, M.L.D. Battery-supercapacitor hybrid energy storage system in standalone DC microgrids: A review. *IET Renew. Power Gener.* **2016**, *11*, 461–469. [CrossRef]
27. Kinhekar, N.; Padhy, N.P.; Li, F.; Gupta, H.O. Utility oriented demand side management using smart AC and micro DC grid cooperative. *IEEE Trans. Power Syst.* **2016**, *31*, 1151–1160. [CrossRef]
28. Gabbar, H.A.; Othman, A.M. Performance optimisation for novel green plug-energy economizer in micro-grids based on recent heuristic algorithm. *IET Gener. Transm. Distrib.* **2016**, *10*, 678–687. [CrossRef]
29. Ackermann, T.; Cherevatskiy, S.; Brown, T.; Eriksson, R.; Samadi, A.; Ghandhari, M.; Söder, L.; Lindenberger, D.; Jägemann, C.; Hagspiel, S.; et al. Smart Modeling of Optimal Integration of High Penetration of PV-SMOOTH PV. Available online: http://smooth-pv.info/doc/SmoothPV_Final_Report_Part1.pdf. (accessed on 19 January 2018).
30. Arboleya, P.; Gonzalez-Moran, C.; Coto, M.; Falvo, M.C.; Martirano, L.; Sbordone, D.; Bertini, I.; Di Pietra, B. Efficient energy management in smart micro-grids: ZERO grid impact buildings. *IEEE Trans. Smart Grid* **2015**, *6*, 1055–1063. [CrossRef]

31. Chub, A.; Husev, O.; Blinov, A.; Vinnikov, D. Novel Isolated Power Conditioning Unit for Micro Wind Turbine Applications. *IEEE Trans. Ind. Electron.* **2017**, *64*, 5984–5993. [CrossRef]
32. RRahman, S.A.; Varma, R.K.; Vanderheide, T. Generalised model of a photovoltaic panel. *IET Renew. Power Gener.* **2014**, *8*, 217–229. [CrossRef]
33. Silva, E.A.; Bradaschia, F.; Cavalcanti, M.C.; Nascimento, A.J. Nascimento. Parameter estimation method to improve the accuracy of photovoltaic electrical model. *IEEE J. Photovolt.* **2016**, *6*, 278–285. [CrossRef]
34. Bhayo, M.A.; Yatim, A.H.M.; Khokhar, S.; Aziz, M.J.A.; Idris, N.R.N. Modeling of Wind Turbine Simulator for analysis of the wind energy conversion system using MATLAB/Simulink. In Proceedings of the IEEE Conference on Energy Conversion (CENCON), Johor Bahrum, Malaysia, 19–20 October 2015; pp. 122–127.
35. Breaz, E.; Gao, F.; Blunier, B.; Tirnovan, R. Mathematical modeling of proton exchange membrane fuel cell with integrated humidifier for mobile applications. In Proceedings of the IEEE Transportation Electrification Conference and Expo (ITEC), Dearborn, MI, USA, 18–20 June 2012; pp. 1–6.
36. Mihet-Popa, L.; Boldea, I. Dynamics of control strategies for wind turbine applications. In Proceedings of the 10th International Conference on Optimisation of Electrical and Electronic Equipment, OPTIM 2006, Poiana Brasov, Romania, 18–19 May 2006; pp. 199–206.
37. Yu, S.Y.; Kim, H.J.; Kim, J.H.; Han, B.M. SoC-based output voltage control for BESS with a lithium-ion battery in a stand-alone DC microgrid. *Energies* **2016**, *9*, 924. [CrossRef]
38. Zhou, N.; Liu, N.; Zhang, J.; Lei, J. Multi-Objective optimal sizing for BATTERY Storage of PV-based microgrid with demand response. *Energies* **2016**, *9*, 591. [CrossRef]
39. Mihet-Popa, L.; Camacho, O.M.F.; Norgard, P.B. Charging and discharging tests for obtaining an accurate dynamic electro-thermal model of high power lithium-ion pack system for hybrid and EV applications. In Proceedings of the IEEE PES Power Tech Conference, Grenoble, France, 16–20 June 2013.
40. Camacho, O.M.F.; Mihet-Popa, L. Fast Charging and Smart Charging Tests for Electric Vehicles Batteries using Renewable Energy. *Oil Gas Sci. Technol.* **2016**, *71*, 13. [CrossRef]

MDPI
St. Alban-Anlage 66
4052 Basel
Switzerland
Tel. +41 61 683 77 34
Fax +41 61 302 89 18
www.mdpi.com

Applied Sciences Editorial Office
E-mail: applsci@mdpi.com
www.mdpi.com/journal/applsci

www.ingramcontent.com/pod-product-compliance
Lightning Source LLC
Chambersburg PA
CBHW061620210926
51850CB00009B/2389
9783036519753